Studies in Computational Intelligence

Volume 1257

The series "Studies in Computational Intelligence" (SCI) publishes new developments and advances in the various areas of computational intelligence—quickly and with a high quality. The intent is to cover the theory, applications, and design methods of computational intelligence, as embedded in the fields of engineering, computer science, physics and life sciences, as well as the methodologies behind them. The series contains monographs, lecture notes and edited volumes in computational intelligence spanning the areas of neural networks, connectionist systems, genetic algorithms, evolutionary computation, artificial intelligence, cellular automata, self-organizing systems, soft computing, fuzzy systems, and hybrid intelligent systems. Of particular value to both the contributors and the readership are the short publication timeframe and the world-wide distribution, which enable both wide and rapid dissemination of research output.

Indexed by SCOPUS, DBLP, WTI AG (Switzerland), zbMATH, SCImago.

All books published in the series are submitted for consideration in Web of Science.

Soumi Majumder · Nilanjan Dey

AI-Driven Wellbeing to Enhance Lives and Work Environments

Springer

Soumi Majumder
Future Institute of Engineering
and Management
Future Business School
Kolkata, West Bengal, India

Nilanjan Dey
Department of Computer Science
and Engineering
Techno International New Town
Kolkata, West Bengal, India

ISSN 1860-949X ISSN 1860-9503 (electronic)
Studies in Computational Intelligence
ISBN 978-3-032-15734-8 ISBN 978-3-032-15735-5 (eBook)
https://doi.org/10.1007/978-3-032-15735-5

The original submitted manuscript has been translated into English. The translation was done using artificial intelligence. A subsequent revision was performed by the author(s) to further refine the work and to ensure that the translation is appropriate concerning content and scientific correctness. It may, however, read stylistically different from a conventional translation.

This Springer imprint is published by the registered company Springer Nature Switzerland AG
The registered company address is: Gewerbestrasse 11, 6330 Cham, Switzerland

Preface

In an era where technology is swiftly transforming the essence of our everyday existence, the convergence of artificial intelligence and human health has become increasingly vital. *AI-Driven Wellbeing to Enhance Lives and Work Environments* examines the potential of intelligent systems to be utilized not only for productivity and efficiency but also for fostering healthier, more content individuals and communities. This book emerges from a strong conviction that AI, when carefully crafted and compassionately applied, can act as a significant partner, enhancing our workplaces, aiding mental health, and fostering settings where individuals genuinely flourish. Emphasizing mind-body health enables people to lead more harmonious and satisfying lives. It assists individuals in handling stress, developing resilience, and attaining a feeling of inner peace. The body and mind are strong partners. Chronic stress or negative feelings may result in physical issues such as muscle tension or sickness, whereas health problems can influence mood and mental sharpness. Mind-body wellness is crucial, as it fosters a comprehensive, empowered, and harmonious approach to health, acknowledging that our thoughts, emotions, and actions are all linked and important to our complete well-being. This work seeks to motivate leaders, innovators, and ordinary readers through research, actual case studies, and future-focused insights to utilize AI as a catalyst for beneficial change in personal and professional realms. In the book, Chap. 1 covers Mind-Body Wellbeing; Chap. 2 depicts AI for Work-Life Harmony; Chap. 3 presents AI technologies for Driving Wellness Transformation; AI in Workplace Wellness Programs is discussed in Chap. 4; Chap. 5 describes AI applications for general well-being; AI and organizational wellness culture are discussed in Chap. 6; Chap. 7 reports the ethical considerations and challenges; and future trends in AI for holistic wellness are covered in Chap. 8. Case studies and global best practices, along with the conclusion of the book, are discussed in Chaps. 9 and 10, respectively. This book focuses on AI-powered platforms that can offer real-time mental health assistance, identify initial signs of distress, and link individuals to suitable resources. These technologies provide privacy, accessibility, and immediate interventions that enhance conventional support systems. The book emphasized the need to create AI solutions that focus on user privacy, consent, and inclusivity. It supports clear algorithms and continuous human supervision, ensuring that technological progress enhances rather than substitutes human judgment and compassion.

Kolkata, India

Soumi Majumder
Nilanjan Dey

Contents

About the Authors

Soumi Majumder is a Research Scholar in the Department of Business and Management at Lincoln University College, Malaysia, and an Assistant Professor at the Future Business School, Future Institute of Engineering and Management, Kolkata, India. She has over 10 years of academic and 2 years of industry experience and has previously served at institutions such as Sister Nivedita University, Techno India College of Technology, and NSHM College of Management and Technology. She has authored five books, including titles published by Springer Nature, Taylor & Francis, and Emerald, along with two edited volumes. Her research focuses on quality of work life, stress management, employee engagement, leadership, and entrepreneurship, with over 25 papers in reputed national and international journals and conferences. She holds one patent and four copyrights from the Government of India and serves as a reviewer for journals by Elsevier, Springer, Emerald, and Sage.

Nilanjan Dey received the B.Tech. and M.Tech. in Information Technology from West Bengal Board of Technical University and the Ph.D. in Electronics and Telecommunication Engineering from Jadavpur University, Kolkata, India, in 2005, 2011, and 2015. Currently, he is a professor with Techno International New Town, Kolkata, and a visiting fellow of the University of Reading, UK. He is the Editor-in-Chief of the *International Journal of Ambient Computing and Intelligence*, Associate Editor of *IEEE Transactions on Technology and Society*, and series Co-editor of *Springer Tracts in Nature-Inspired Computing and Data-Intensive Research* (Springer Nature) and *Advances in Ubiquitous Sensing Applications for Healthcare* (Elsevier). He is also on the Editorial Board of *IEEE Data Descriptions*. He is a Fellow of IETE, Senior Member, IEEE and member of IE, ISOC, etc.

Chapter 1
Understanding Mind–Body Wellbeing

1.1 Defining Mind–Body Wellness

Mind–body wellness refers to the integrated idea that the mind and body are closely linked and that attaining ideal health requires supporting both mental and physical well-being simultaneously [1]. This method acknowledges that thoughts, feelings, and attitudes can greatly affect physical health, just as physical health can influence mental and emotional well-being. For example, prolonged stress or pessimistic thoughts can cause physical issues such as muscle tightness or sickness, whereas bodily problems can impact mood and cognitive sharpness.

Mind–body wellness is essential, as it acknowledges the profound connection between mental and physical health, emphasizing that both need to be cared for to achieve overall well-being [2]. Wellness is not a passive condition; it demands deliberate actions and routines that promote both mental and physical well-being. This involves making intentional decisions daily to nurture one's body and mind instead of believing that wellness will occur on its own. Mind–body wellness involves the collaboration of numerous bodily systems. Focusing on both mental and physical elements leads to enhancements that are more lasting and significant than targeting only one aspect. Emphasizing mind–body health enables people to lead more harmonious and satisfying lives. It assists individuals in handling stress, developing resilience, and attaining a feeling of inner peace. The body and mind are strong partners. Chronic stress or negative feelings may result in physical issues such as muscle tension or sickness, whereas health problems can influence mood and mental sharpness [3]. Mind–body wellness is crucial, as it fosters a comprehensive, empowered, and harmonious approach to health, acknowledging that our thoughts, emotions, and actions are all linked and important to our complete well-being.

S. Majumder and N. Dey, *AI-Driven Wellbeing to Enhance Lives and Work Environments*, Studies in Computational Intelligence 1257,
https://doi.org/10.1007/978-3-032-15735-5_1

1.2 Physical, Mental, and Emotional Wellness Components

Mind–body wellness is a holistic approach to health that acknowledges the relationships among physical, mental, and emotional aspects to achieve complete well-being [4]. It highlights methods that encourage balance between the mind and body, including mindfulness, physical activity, and stress reduction, to enhance overall well-being. Elements of mind–body wellness are discussed below.

I. **Physical Wellness**

Physical wellness involves looking after one's body to support optimal health and overall well-being. It includes multiple elements, such as physical exercise, nutritious eating, adequate rest, and stress management. Figure 1.1 is the representation of the dimension of wellness that consists of factors, i.e., emotional, ecological, financial, intellectual, occupational, physical, social, and spiritual.

It also entails making intentional decisions to avoid sickness and harm. Fundamentally, it involves making a conscious effort to achieve a healthy body and way of life. It emphasizes maintaining a healthy body via proper nutrition, consistent exercise, sufficient sleep, and regular medical check-ups. Keywords include (i) exercise: engaging in activities such as yoga, tai chi, or cardiovascular workouts that improve strength, flexibility, and heart health while also fostering mental clarity; (ii) nutrition: eating balanced, nutrient-dense foods to energize the body and enhance cognitive capabilities; (iii) sleep: empasizing restorative sleep to support physical healing and emotional stability; (iv) preventive care: consistent health screenings and positive habits to avert illness and sustain bodily functions; and (v) mind–body

Fig. 1.1 Dimensions of wellness

connection: physical practices such as yoga not only increase fitness but also alleviate stress and improve mental concentration through mindful movement [5].

II. **Mental Wellness**

Mental wellness, referred to as mental well-being, represents a condition of favourable mental health that includes emotional, psychological, and social well-being. It is not only the lack of mental illness; it is about thriving and positively influencing the life and the lives of those around people. Mental well-being includes managing life difficulties, nurturing positive relationships, and engaging with one's community. It encompasses cognitive wellness, intellectual engagement, and stress reduction to maintain clarity, concentration, and resilience [6].

III. **Emotional Wellness**

Emotional wellness encompasses acceptance, awareness, and healthy management of emotions. The National Center for Emotional Wellness defines emotional wellness as "awareness, understanding, and acceptance of our feelings and the ability to navigate challenges and change [7]. It centers on understanding, managing, and expressing emotions in a healthy way, fostering resilience and positive relationships. The key aspects are (i) emotional awareness, i.e., recognizing and processing emotions through self-reflection or therapy; (ii) positive coping mechanisms, i.e., using strategies such as gratitude journaling or social support to navigate challenges; (iii) relationship building, i.e., cultivating meaningful connections that provide emotional support and reduce feelings of isolation; and (iv) connection to the mind–body, i.e., emotional health influences physical health (e.g., chronic stress can lead to heart disease) and mental clarity, whereas practices such as mindfulness can stabilize emotions.

IV. **Holistic Integration**

Holistic integration involves combining all the elements of something (as an individual, an organization, or a system) to form a cohesive, interrelated entirety. This highlights that the entirety exceeds the sum of its components and that achieving optimal performance or success necessitates addressing and aligning all the factors involved. Mind–body wellness acknowledges that these elements rely on each other [8]. For example, engaging in physical activity increases endorphin levels, improving emotional health and mental sharpness. Mental exercise, such as meditation, can reduce blood pressure and increase emotional strength. Emotional equilibrium enhances the drive for exercise and lessens mental exhaustion. By engaging all three elements through activities such as yoga, conscious eating, or counselling, mind–body wellness fosters a harmonious, healthy lifestyle where each part supports the others [4, 9].

1.3 Role of Technology in Wellness

Technology greatly influences wellness by improving fitness monitoring, encouraging mental health assistance, and facilitating personalized, accessible healthcare via tools such as telemedicine and AI-powered solutions. Wearable devices, smartphone applications, and AI systems are changing how people monitor their health and wellness, allowing for a more proactive and customized approach to their needs [3]. In the modern world, technology effortlessly blends into our everyday activities, influencing every facet of our existence—the workplace included. The business environment has experienced significant changes driven by technology, altering not only our work methods but also our approach to valuing and prioritizing employee well-being. The effect of technological progress on employee wellness initiatives is clear, leading to a substantial change in how companies view and prioritize the health of their staff. The significant impact of technology on people's physical, mental, and social health is prompting numerous businesses to adopt creative approaches to enhance their wellness programs. Figure 1.2 depicts AI in diagnosis. It includes two aspects: early detection of mental health disorders and predictive models. Early detection consists of analysing electronic health records and analysing speech, text, and facial cues. Predictive models represent a multifunctional predictive model for improvement in treatment. In contrast to conventional wellness programs, which often necessitate physical presence at designated sites or compliance with strict timetables, technology enables employees to participate in wellness activities from almost any location. Fitness applications, online tools, and virtual coaching services offer employees the convenience of selecting when and where they take part, aligning health initiatives with their hectic routines and varied lifestyles. Employees who participate in wellness activities tend to enjoy enhanced physical and mental health. Consequently, this results in decreased levels of absenteeism and presenteeism. A Gallup study indicated that companies with highly engaged employees experience 81% variance in absenteeism and a 14% rise in productivity [10].

Technology plays a significant role in enhancing wellness by providing tools, platforms, and data-driven solutions that support physical, mental, and emotional well-being. Its influence extends to multiple aspects of well-being, making wellness more attainable, individualized, and effective [11]. The impact of technology on wellness is as follows:

1.3.1 Impact of Technology on Physical Wellness

I. Wearable Gadgets and Activity Monitors: Gadgets (e.g., Fitbit, Apple Watch), and Garmin monitor metrics, including steps, heart rate, sleep patterns, and calories expended. These resources assist individuals in tracking their exercise, establishing fitness objectives, and upholding responsibility [12].

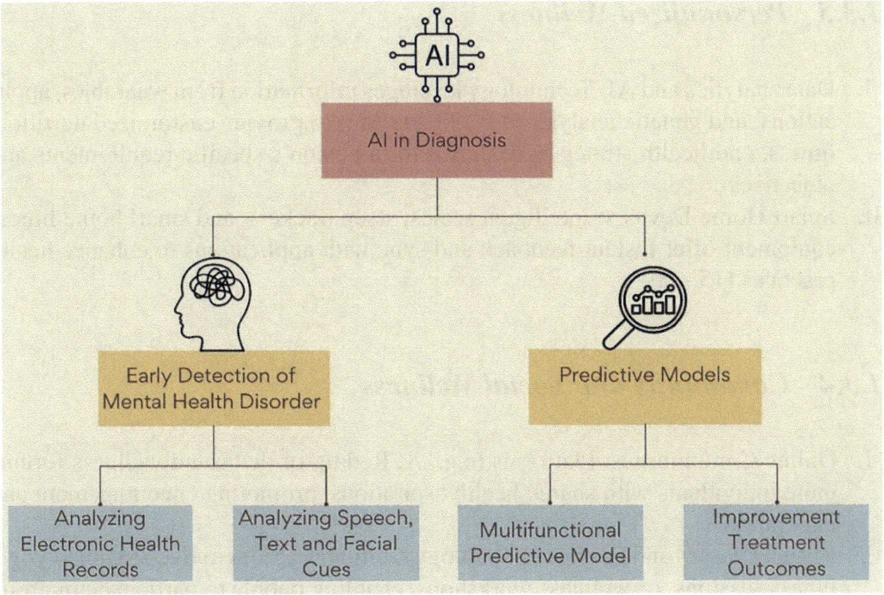

Fig. 1.2 AI in diagnosis

II. Health Apps: Applications such as MyFitnessPal, Strava, or the Nike training club offer exercise regimens, dietary monitoring, and community assistance, helping users enjoy healthier living.
III. Telemedicine and Remote Monitoring: Telehealth platforms facilitate consultations with medical professionals from home, whereas remote monitoring tools (for glucose monitors or blood pressure cuffs) support real-time health management, particularly for ongoing conditions [13].

1.3.2 Impact of Technology on Mental and Emotional Wellness

I. Mental Health Apps: Applications such as Headspace, Calm, and BetterHelp provide guided meditations, mindfulness activities, and therapist access, assisting users in handling stress, anxiety, and depression.
II. AI-Powered Support: Chatbots and AI applications (e.g., Woebot), deliver cognitive behavioral therapy (CBT) methods and emotional assistance, making mental health resources easily available [14].
III. Virtual reality (VR) therapy: VR aids in mental health treatment by providing exposure therapy, reducing stress, and creating immersive relaxation experiences.

1.3.3 Personalized Wellness

I. Data analytics and AI: Technology leverages information from wearables, applications, and genetic analysis (e.g., 23 and Me) to provide customized nutrition, fitness, and health strategies designed for a person's specific requirements and objectives.
II. Smart Home Devices: Intelligent scales, sleep trackers, and smart home fitness equipment offer instant feedback and sync with applications to enhance health practices [15, 16].

1.3.4 Community and Social Wellness

I. Online Communities: Platforms (e.g., X, Reddit, or dedicated wellness forums unite individuals with shared health aspirations, promoting encouragement and inspiration.
II. Virtual Classes and Events: Technology facilitates access to live-streamed yoga, fitness sessions, or wellness workshops, enabling people to participate in group activities from a distance [17].

1.3.5 Access to Information and Education

I. Health Portals and blogs: Websites and apps provide evidence-based information on nutrition, exercise, and mental health, empowering individuals to make informed decisions.
II. Wearable Feedback: Instantaneous information from wearables informs users about their health trends, promoting proactive adjustments.

1.3.6 Workplace Wellness

I. Wellness Platforms: Platforms such as Virgin Pulses or wellable connections with corporate systems to increase employee well-being via challenges, stress reduction initiatives, and health assessments [18].
II. Ergonomic Tech: Intelligent desks, posture trackers, and blue-light-filtering eyewear enhance workplace ergonomics and reduce physical stress.

1.4 Initiatives for Advanced Technologies Transforming Corporate Wellness

New technologies are changing corporate wellness programs in ways that make them more personalized, accessible, and effective than ever before. AI and machine learning systems can evaluate employee health information to provide customized wellness advice, whereas wearable devices track real-time physical activity, sleep habits, and stress levels, enabling individuals to control their health. Virtual reality (VR) is utilized to aid immersive mindfulness and stress management practices, whereas mobile health applications offer immediate coaching, resources, and prompts to promote healthy behaviors. Furthermore, data analytics tools assist organizations in recognizing trends and assessing the impact of wellness initiatives, fostering ongoing enhancement. Through the use of these innovative tools, organizations are developing flexible, responsive wellness approaches that benefit individual health and overall company success. A few advanced technologies are discussed in this section.

I. Wearable Devices

By integrating wearable devices and health applications, workers can track their health metrics as they happen [19]. This information enhances self-awareness, enabling the tailoring of wellness plans to personal needs and preferences. Workers can obtain personalized workout plans, dietary advice, and stress relief techniques, ensuring that the wellness initiative matches their health objectives and addresses particular issues of interest (Fig. 1.3).

II. Mobile Applications

The mobility of mobile applications allows employees to embrace healthy habits, track their diet, and monitor their stress levels. Often, mobile applications provide

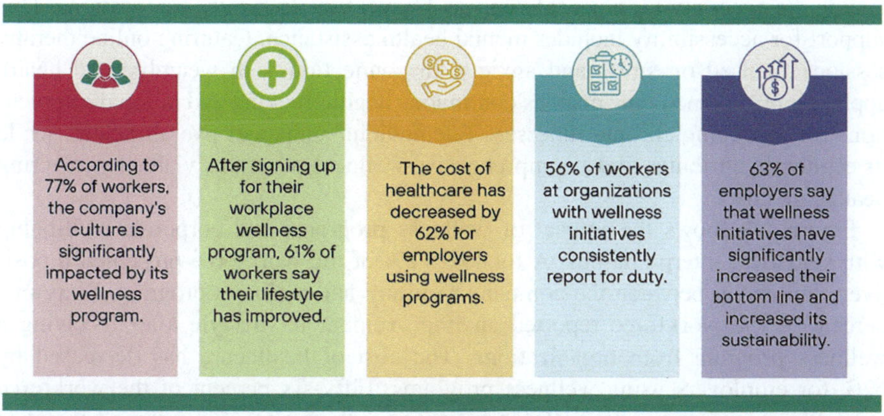

Fig. 1.3 Impact of the wellness program and corporate well-being

mindfulness exercises that can greatly improve mental health. Mobile apps impact people and promote a culture of wellness in the workplace by offering tailored information, helpful reminders, and guidance for sustaining a healthy lifestyle [20].

III. Online health assessments

Online health evaluations provide employees with a detailed and tailored review of their health condition. When data on different health metrics, lifestyle habits, and risk factors are collected, these evaluations offer a baseline comprehension that establishes a basis for focused wellness initiatives [21].

IV. Virtual Wellness Classes

Ranging from fitness workouts and yoga classes to meditation and nutrition workshops, online wellness sessions address multiple facets of holistic health, offering employees accessible chances to focus on their well-being. Live sessions frequently offer chances for immediate engagement with teachers and other attendees, forming a digital environment for collective wellness activities. This feeling of companionship not only enhances motivation but also fosters a positive and encouraging workplace culture, even in a remote or hybrid setting.

V. Gamification

Technology incorporates gamification into wellness programs, boosting engagement and making the entire experience more enjoyable. Game-like challenges, online contests, and incentive programs not only encourage employees to engage regularly but also introduce a sense of enjoyment to the wellness experience [22]. This method transforms wellness activities from a task into an exciting experience, fostering ongoing long-term participation.

VI. Telemedicine

The incorporation of telemedicine into corporate wellness initiatives allows employees to interact with healthcare professionals via virtual appointments, minimizing the necessity for physical visits and travel time to healthcare locations. This support for accessibility includes mental health assistance, featuring online therapy sessions focused on stress and anxiety. In connection with wearable and health applications, telemedicine enables continuous health tracking and tailored interventions for managing chronic illnesses. Telemedicine enhances overall health, while its economic attributes make comprehensive wellness programs viable by lowering healthcare costs.

Figure 1.3 shows the impact of wellness programs and corporate well-being with statistical interpretations. A total of 77% of the workforce presented a positive relationship between the company's culture and wellness program. Sixty-one percent of the workforce reported an improvement in lifestyle after receiving a wellness program from organizations. The cost of healthcare has decreased by 62% for employers using wellness programs. Fifty-six percent of the workforce in healthcare is happy with wellness initiatives at their work. Sixty-three percent of employers say that wellness initiatives have significantly increased their bottom line and increased their sustainability (source: https://www.collidu.com/presentation-corporate-wellness-program).

VII. Health Management Platforms

Health management platforms provide employees with health insights and goal tracking to take charge of their wellness journey. Health management platforms allow employers to optimize wellness programs by using data analytics to evaluate participation rates, assess the effectiveness of initiatives, and customize programs to meet the unique needs of their workforce. They can provide tailored and engaging content, enhancing employee involvement and interaction [23].

VIII. Virtual reality

The use of virtual reality technologies creates opportunities for developing engaging and immersive health and wellness experiences (e.g., virtual meditation practices). Through virtual reality, workers can participate in various virtual tasks and challenges aimed at improving their well-being. Employing VR for stress relief offers an entertaining and captivating method for employees to be stressed and rejuvenated, enhancing a comprehensive and enjoyable wellness initiative. Corporate wellness initiatives frequently depend on wearables (e.g., Fitbit, Apple Watch) and applications that gather sensitive information about employees, including health metrics, sleep habits, and mental health indicators. Data leaks or unapproved distributions to external parties (insurers or employers) can diminish trust. A 2023 data breach of a major wellness platform revealed user health information, causing worries about the platform (e.g., X). Workers might be reluctant to engage if they are concerned that their data could influence job stability or insurance costs. Adhering to the GDPR, HIPAA, or CCPA is challenging, particularly for international corporations [24].

Strong encryption, anonymized data handling, and clear data policies are crucial yet expensive to establish. AI-based wellness applications (those suggesting tailored fitness or mental health strategies) rely on algorithms developed from expansive datasets. If the training data are not diverse (e.g., biased towards specific demographics), the suggestions may be ineffective or detrimental to marginalized groups. A 2024 study, for example, noted biases in AI fitness apps that misjudged calorie requirements for non-Western groups. Incorrect suggestions can undermine program efficiency and worsen health disparities among workers. Wellness technologies frequently monitor employee behaviors, such as activity levels or stress indicators, which raises ethical concerns. Employees might view this monitoring as surveillance, particularly if the data are accessible to HR or management. A 2025 X thread emphasized employee resistance to a firm leveraging wearable data for productivity evaluation. A sense of overreach may erode trust and morale, possibly resulting in legal or reputational challenges. Effective communication regarding data utilization, opt-in consent, and firm limits on data sharing is crucial yet difficult to maintain consistently. Although technologies such as telehealth and AI coaching may lower costs over time, the initial setup is costly [25, 26].

Technologies such as AI chatbots or virtual therapy services (e.g., Woebot) offer scalable assistance for mental health but cannot completely replace human interaction. Excessive dependence on technology-based solutions can result in shallow involvement, especially for mental health challenges that need empathy. Workers on X have voiced discontent when AI tools do not meet intricate emotional requirements. Programs can seem impersonal, diminishing their efficacy for employees facing significant wellness issues.

The swift uptake of wellness technologies exceeds the pace of regulatory frameworks, leading to compliance difficulties. The EU's AI Act (2024) and comparable laws categorize health-related AI as high risk, necessitating thorough testing and transparency.

1.5 The Rise of AI in Wellness: Opportunities and Challenges

Healthcare providers, therapists, and researchers are progressively discovering that artificial intelligence (AI) can serve as an effective resource in delivering mental health services. This article shows that an increasing amount of research indicates the potential of AI in diagnosing illnesses, creating treatments, and facilitating more individualized methods and therapies. Chatbots are being utilized more frequently to provide support and communication for mental health patients throughout their treatment [26]. They can assist in managing symptoms and monitoring specific keywords that might prompt a referral and direct interaction with a human mental health professional. AI can also be utilized to examine patient medical records, behavioral information, voice recordings gathered from calls to intervention services, and various other data sources, employing machine learning to identify warning signs of mental issues before they escalate to a critical stage.

Researchers at Vanderbilt University Medical Center also discovered that machine learning could analyse hospital admission data, demographic information, and clinical data to predict suicidal behavior in individuals with 80% accuracy [27]. The Alan Turing Institute is working on a distinct project that uses AI to predict mental health problems. In this study, scientists are investigating how to utilize extensive datasets from people without visible mental health symptoms to forecast which individuals may eventually exhibit symptoms throughout their lives. AI has been utilized to forecast instances where patients are more inclined to benefit from cognitive behavioral therapy (CBT) and thus be less prone to needing medication. A study in JAMA Psychology revealed that deep learning can support the validation of the effectiveness of CBT as a treatment approach, which may lessen the requirement to prescribe medications for certain patients. An alternative project focused on the use of AI to predict mental health issues is underway at the Alan Turing Institute. In this research, researchers are exploring how to leverage large datasets from individuals without apparent mental health symptoms to predict which people might later display symptoms over their lifetime [28]. AI has been used to predict cases in which patients are more likely to benefit from cognitive behavioral therapy (CBT) and consequently are less likely to require medication. As antidepressant and antipsychotic drugs can lead to side effects that might limit life, this could greatly improve outcomes for specific patients. Research published in JAMA Psychology indicated that deep learning can aid in confirming the efficacy of CBT as a treatment method, potentially reducing the need for medication prescriptions for some patients.

1.5.1 Opportunities of AI in Wellness

AI examines extensive datasets—genetic information, lifestyle choices, and health indicators—to provide personalized wellness suggestions. For example, firms such as Viome utilize AI to examine gut microbiome information, offering tailored dietary recommendations to enhance health. Devices such as the Fitbit or Apple Watch utilize AI to monitor sleep, activity, and heart rate, providing users with immediate insights to enhance physical fitness and mental health. Chatbots and virtual therapists powered by AI (e.g., Woebot, Youper) offer accessible support for mental health. These tools employ natural language processing (NLP) to involve users in cognitive behavioral therapy (CBT) or mindfulness practices, enhancing access to mental health care in an affordable and scalable way, particularly in neglected areas. AI models can forecast health risks by examining trends in medical records, lifestyle information, and social determinants of health [29]. AI algorithms in platforms such as Google Health can predict possible ailments with diabetes or heart problems, allowing preventive actions before symptoms develop. Apps powered by AI (e.g., MyFitnessPal, Noom) utilize machine learning to develop personalized meal plans and exercise regimens. These platforms adjust to individual progress, preferences, and objectives, enhancing adherence and results in comparison with standard fitness programs. AI makes wellness accessible by providing affordable, round-the-clock solutions [30]. Online yoga teachers, mindfulness applications (e.g., Calm), and AI-based sleep solutions connect with worldwide audiences, overcoming geographical and financial obstacles. This is especially significant in areas with restricted access to healthcare providers [31].

1.5.2 Challenges of AI in Wellness

Wellness applications gather sensitive information—medical records, biometric data, and behavioral trends—to address concerns about potential breaches or exploitation. Notable events, such as the 2023 data breach, highlight the dangers associated with retaining personal health information. Adhering to regulations such as the GDPR or HIPAA is essential yet intricate [1]. AI systems can reinforce biases when trained on unrepresentative data. If wellness algorithms are created using data mainly from specific demographics, they might offer misleading recommendations for marginalized groups, worsening health inequalities. Relying on AI tools threatens to diminish human connection in wellness care. Although they are beneficial, mental health applications cannot completely replace the understanding and subtlety of human therapists. Excessive use can result in "digital overload," causing users to feel inundated by ongoing data monitoring. If AI-generated wellness suggestions are not thoroughly verified, they can disseminate false information [32]. For example, unproven AI-based dietary supplements or wellness trends advertised on social media platforms (e.g., X) can deceive users.

Although AI can reduce expenses, high-end wellness tools (e.g., sophisticated wearables, subscription apps) continue to be costly, resulting in a digital gap. Rural or low-income communities might not have the necessary devices or internet connectivity to take advantage of AI-based wellness solutions. Certain users support open-source AI models to promote transparency in health applications. The AI market in healthcare, which encompasses wellness, is expected to surpass $100 billion by 2028, fueled by innovations in wearables, telehealth, and predictive analytics (according to industry reports from 2025) [33, 34]. Governments are imposing stricter regulations on AI. The EU's AI Act (2024) categorizes health-related AI as high risk, necessitating rigorous testing and openness, potentially influencing the creation of wellness apps.

AI systems can examine trends in employee actions, interactions, and productivity metrics to detect early indicators of burnout (lower engagement or higher absenteeism). This enables organizations to act before burnout escalates. Applications and platforms utilizing AI can suggest personalized wellness tools (e.g., mindfulness practices), reminders for breaks, or adjustments to workload) designed to suit personal requirements, aiding in the prevention or alleviation of burnout. AI assists managers in balancing workloads by examining task allocation and recommending modifications. This guarantees that no single team member is overwhelmed, lowering the likelihood of burnout [35, 36]. By automating tedious or repetitive tasks, AI allows employees to concentrate on more significant, creative, or strategic activities. This enhances job satisfaction and lessens feelings of fatigue and annoyance. Virtual assistants and chatbots offer immediate assistance and responses, addressing inquiries or assisting workers in handling stress as it occurs.

1.6 Conclusions

This chapter provides an in-depth examination of how artificial intelligence can beneficially impact personal health and work environments. Throughout the chapter, it is evident that AI transcends mere efficiency; it serves as a strong partner in promoting emotional, mental, and physical wellness. Through tailored health advice, stress identification, or improved workplace experiences, AI creates more supportive atmospheres. The integration of AI into well-being presents challenges such as privacy issues, bias in algorithms, and the requirement for digital literacy. This research urges both organizations and individuals to address these issues proactively.

References

1. Rao, T. V. N., Patel, G. S., Bhavana, S., & Swetha, C. (2025). AI-driven decision support system for workplace wellbeing. In *Insights on wellbeing and happiness in the workplace* (pp. 1–18). IGI Global Scientific Publishing.
2. Chawla, N., Ekatpure, V. P., Desai, N., Raje, H., Lenka, R., & Bhandwalkar, V. (2025). AI driven work life integration: How smart technologies transforming the modern workplace. In *AIP conference proceedings* (Vol. 3306, p. 030002). AIP Publishing LLC.
3. Oladele, O. K. (2023). Data-driven work culture: How AI enhances employee well-being and reduces occupational stress.
4. Tariq, M. U. (2025). AI and work–life balance: Transforming employee wellbeing in the modern workplace. In *Strengthening human relations in organizations with AI* (pp. 85–110). IGI Global Scientific Publishing.
5. Majumder, S., Chowdhury, S., Dey, N., & Santosh, K. C. (2022). Balance your work–life: Personal interactive web-interface.
6. Akter, F., Shakil, M. R. U., Rashid, A. A., Fatema, K., Akter, Y., Saky, S. I., et al. (2024). AI-driven workload optimization: Enhancing employee well-being and productivity to promote sustainable economic growth (SDG 8) in Malaysia. *International Journal of Research and Innovation in Social Science, 15*, 1134.
7. Pandey, A., Mishra, S., & Kadanga, A. (2025). Role of artificial intelligence in enhancing employee well. *Global Work Arrangements and Outsourcing in the Age of AI, 25*, 11334.
8. Sanni, B. (2023). An integrated framework for mitigating IT industry burnout: Leveraging AI-driven work pattern monitoring to enhance workforce wellbeing.
9. Mantello, P., & Ho, M. T. (2024). Emotional AI and the future of wellbeing in the postpandemic workplace. *AI and Society, 39*(4), 1883–1889.
10. Tiwari, R., Babu, N. S., Marda, K., Mishra, A., Bhattar, S., & Ahluwalia, A. (2024). The impact of artificial intelligence in the workplace and its effect on the digital wellbeing of employees. *International Journal of Progressive Research in Engineering Management and Science (IJPREMS), 4*(6), 2422–2427.
11. Jain, C., & Kanwar, M. J. (2024). The AI-driven workplace: How automation is reshaping flexible work arrangements.
12. Banerjee, S., Agarwal, A., & Bar, A. K. (2024). Securing well-being: Exploring security protocols and mitigating risks in ai-driven mental health chatbots for employees. *American Journal of Computer Science and Technology, 7*(1), 1–8.
13. Wadhwa, S. N., Bhardwaj, G., Srivastava, A. P., & Malik, R. (2025). AI-driven job insecurity and work performance: Unveiling the mediating role of psychological well-being. *International Journal of Information Technology, 14*, 1–12.
14. Dhand, S. (2025). Human-centric digital workplace: Designing for employee wellbeing and productivity. In *Human-centric, sustainable, and resilient organizations in the digital age* (pp. 125–146). IGI Global Scientific Publishing.
15. Kaur, J. (2018). Promoting well-being in AI workplaces: A Wi-Fi-enabled, sentiment analysis approach to support women.
16. Parsakia, K., & Tabar, S. H. S. A. (2024). The future of occupational health: Anticipating risks in the evolving workplace. *Journal of Foresight and Health Governance, 1*(2), 30–45.
17. Ghosh, S., Majumder, S., & Peng, S. L. (2023). An empirical study on adoption of artificial intelligence in human resource management. In *Artificial intelligence techniques in human resource management* (pp. 29–85). Apple Academic Press.

18. Saeidnia, H. R., Hashemi Fotami, S. G., Lund, B., & Ghiasi, N. (2024). Ethical considerations in artificial intelligence interventions for mental health and well-being: Ensuring responsible implementation and impact. *Social Sciences, 13*(7), 381.
19. Lee, J. (2025). Smart solutions for student well-being: AI-driven mental health support in a polytechnic. In *The rise of intelligent machines* (pp. 211–242). Chapman and Hall/CRC.
20. Das, S., Chakraborty, S., Sajjan, G., Majumder, S., Dey, N., & Tavares, J. M. R. (2022). Explainable AI for predictive analytics on employee attrition. In *International conference on soft computing and its engineering applications* (pp. 147–157). Springer.
21. Valarmathi, R., Yazhini, G., Narmadha, L. S., & Janani, A. M. (2024). Enhancing mental well-being through AI-driven virtual environments: The Merrytopia project. In *Proceedings of the 2024 international conference on power, energy, control and transmission systems (ICPECTS)* (pp. 1–6). IEEE.
22. Revathi, K., Priyanka, S., Priyadharshini, S., Niranjan, N. N., Visalachi, N., & Sumathi, P. (2025). AI-driven approaches to enhancing mental wellbeing and stress relief. In *Proceedings of the 2025 international conference on multi-agent systems for collaborative intelligence (ICMSCI)* (pp. 925–931). IEEE.
23. Ahumada-Tello, E., Evans, R. D., Romero-Gómez, D., López-García, J., & Castañón-Puga, M. (2023). Impact of AI on employee well-being and decision-making: Insights from OECD member countries. In *Proceedings of the 2023 IEEE global conference on artificial intelligence and internet of things (GCAIoT)* (pp. 121–126). IEEE.
24. Yoon, S., Goh, H., Low, X. C., Weng, J. H., & Heaukulani, C. (2024). User perceptions and utilization of features of an AI-enabled workplace digital mental wellness platform 'mindline at work.' *BMJ Health and Care Informatics, 31*(1), e101045.
25. Rathee, R., & Malik, S. (2024). Exploring AI-based machine learning applications in leadership for enhancing employee happiness and wellbeing. In *Proceedings of the 2024 international conference on advances in computing, communication and applied informatics (ACCAI)* (pp. 1–6). IEEE.
26. Jean, G. (2023). Redefining work–life balance in IT: The power of AI-driven work pattern adjustments.
27. Singh, B. (2025). Assimilating employee well-being into sustainable business practices for successful organization: A path to long-term growth and productivity. In *Prioritizing employee mental health and well-being for organizational success* (pp. 161--184). IGI Global Scientific Publishing.
28. Joseph, S. (2024). Organizational workforce management in the digital age: The role of technocultural interventions in mitigating the negative impacts of AI-driven technological change. *Archives of Current Research International, 24*(10), 10–9734.
29. Oluwagbade, E. (2025). Ethical AI in employee wellness: Balancing data-driven insights with employee privacy and trust.
30. Shinde, G. R., Majumder, S., Bhapkar, H. R., & Mahalle, P. N. (2022). *Quality of work–life during pandemic*. Springer.
31. Zhai, S., Zhang, S., Rong, Y., & Rong, G. (2025). Technology-driven support: Exploring the impact of artificial intelligence on mental health in higher education. *Education and Information Technologies, 64*, 1–29.
32. Akter, F., Shakil, M. R. U., Rashid, A. A., Fatema, K., Akter, Y., Saky, S. I., et al. (2024). AI-driven workload optimization: Enhancing employee well-being and productivity to promote sustainable economic growth (SDG 8) in Malaysia. *International Journal of Research and Innovation in Social Science, 64*, 47772.
33. Gull, A., Ashfaq, J., & Aslam, M. (2023). AI in the workplace: Uncovering its impact on employee well-being and the role of cognitive job insecurity. *International Journal of Business and Economic Affairs, 8*(4), 79–91.
34. Bubicz, M., & Ferasso, M. (2024). Advancing corporate social responsibility in AI-driven human resources management: A maturity model approach. In *Proceedings of the international conference on AI research*. Academic Conferences and publishing limited.

35. Naik, K., Ghosh, S., Soni, H., & Singh, S. (2025). AI-driven burnout detection and eco-volunteering: A sustainable employee well-being model. *International Journal of Environmental Sciences, 11*(9), 697–702.
36. Agarwal, H. (2025). Artificial intelligence in employee well-being and human resource management. *OPJU Business Review, 4*(1), 91–101.

Chapter 2
AI for Work–Life Harmony

AI for work–life balance involves leveraging artificial intelligence (AI) tools to assist people in more effectively managing their professional and personal lives, thereby decreasing stress and enhancing overall wellness. AI tools can streamline repetitive tasks, enhance scheduling, and deliver insights into work behaviors, enabling individuals to concentrate on more significant tasks and enjoy greater leisure time. AI can streamline tedious and time-intensive tasks such as sorting emails, organizing schedules, inputting data, and creating reports, allowing employees to concentrate on more valuable activities. AI can also help establish boundaries by automatically muting notifications during personal time or focused work periods, ensuring that work does not interfere with personal life [1]. Certain AI systems examine work habits and offer tailored suggestions for enhancing concentration, minimizing stress, and maximizing time efficiency. AI for work–life balance focuses on utilizing technology to foster a more rewarding and harmonious work environment, enabling individuals to increase their productivity without increasing their effort and emphasizing their health and happiness alongside their career ambitions. AI is essential for greatly improving the efficiency and accuracy of intricate office relocations, turning what might be a logistical challenge into a smooth operation.

2.1 Intelligent Time Management and Task Automation

Smart time management and task automation involve leveraging artificial intelligence (AI) and machine learning (ML) to increase time utilization and automate routine tasks. This method transcends conventional automation by integrating AI's capacity to learn, adjust, and enhance its performance progressively, resulting in more efficient and effective management of time and task execution [2]. Intelligent automation (IA), also known as cognitive automation, refers to the application of automation technologies—artificial intelligence (AI), business process management

S. Majumder and N. Dey, *AI-Driven Wellbeing to Enhance Lives and Work Environments*, Studies in Computational Intelligence 1257,
https://doi.org/10.1007/978-3-032-15735-5_2

(BPM), and robotic process automation (RPA)—to enhance and expand decision-making throughout organizations. Smart automation streamlines workflows, liberates resources, and enhances operational efficiency through diverse applications. An automotive company could utilize IA to accelerate manufacturing or minimize the chances of human mistakes. A pharmaceutical or life sciences firm can implement intelligent automation to lower expenses and achieve resource efficiencies in areas with repetitive tasks. An insurance company can leverage intelligent automation to determine payments, assess rates, and fulfil compliance requirements. Through the application of machine learning and intricate algorithms to assess both structured and unstructured data, companies can create a knowledge base and generate forecasts derived from that data. This serves as the decision engine for AI. The next element of smart automation is business process management (BPM), which is also referred to as business workflow automation [2]. Business process management streamlines workflows to increase agility and uniformity in business processes. Business process management is utilized in various sectors to optimize processes and enhance interactions and engagement. The third element of IA is robotic process automation (RPA). Robotic process automation employs software robots (or bots) to perform back-office activities, such as data extraction or form completion. Implementing automation in systems and processes while utilizing data analysis to ensure precision can speed up production [3]. AI also allows businesses to grow rapidly without elevating risk, sacrificing quality, or overburdening current employees. Business executives enjoy the advantages of this ability through increased production outputs and enhanced ROI. This integration results in a revolutionary solution that enhances processes and eases workflows to ultimately increase the customer experience. Intelligent automation optimizes processes that were previously made up of manual tasks or rely on outdated systems, which can be resource heavy, expensive, and susceptible to human error. The uses of AI extend throughout sectors, offering efficiencies in various aspects of the company. The automotive sector is significantly influenced by the advancements that producers can achieve through intelligent automation. Through AI, producers can more efficiently forecast and modify production to adapt to fluctuations in supply and demand. They can enhance workflows to increase efficiency and minimize the chances of mistakes in production, support, procurement, and additional fields [4]. The use of robots can decrease reliance on manual labor and enhance defect detection, offering customers a superior product at a reduced cost to the company. A German Volkswagen engine manufacturing facility employs "collaborative robots" that assist the production team with a demanding task in the assembly of engines. Through AI, the insurance sector can nearly eliminate the necessity for manual rate computations or payments and can streamline paperwork handling, such as claims and assessments. Smart automation assists insurance firms in following compliance regulations more effortlessly by verifying that standards are fulfilled. In this way, they can assess the risk of a person or organization and determine a suitable insurance premium (Fig. 2.1).

This section covers the characteristics of time tracking software. Time tracking is crucial for a manager to recognize how time is allocated across various tasks and projects. This is where the time tracking capability of software is useful. It allows

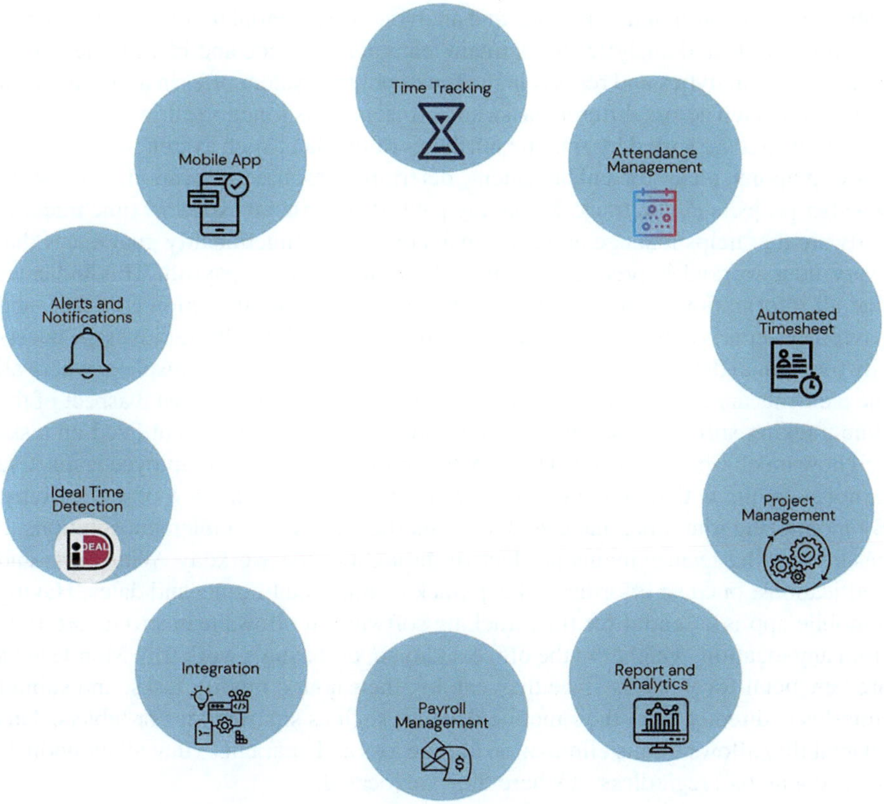

Fig. 2.1 Features of the time tracking software

tracking the hours the team dedicates to different tasks, aiding in understanding the project's efficiency and advancement. This functionality helps monitor the duration allocated to each task, aiding in the management of workloads and deadlines. Employee attendance tracking is one aspect of the time tracking system that simplifies tracking staff attendance. This functionality aids in monitoring when employees clock in and out, thereby assisting in the organization of the work schedule and the recognition of all staff members [5]. Providing real-time attendance data helps spot trends and address issues such as tardiness or absenteeism before they worsen. Effective project management is essential for coordinating all activities and ensuring that the team meets deadlines. The project management function in time tracking software allows for overseeing every element of a project from its start to its finish. It helps assign tasks, set deadlines, manage budgets, oversee client relations, and monitor the project's progress instantly. This functionality allows us to see who is assigned to each task and how much time they have dedicated to it. Integrating project management with time tracking techniques simplifies work management and enhances decision-making. Reports and analytics are essential for obtaining comprehensive reports and analytics to maintain team performance and identify necessary actions. The analytics and reports in the time tracking system offer insights into how time is allocated across different tasks and projects. This function allows the creation of reports that are sorted by specific attributes of the data, such as trends. These data allow pinpoint areas for enhancement, determine resource allocation, and assess whether projects are on track. Managing payroll is a crucial aspect of time tracking software that helps manage employee payments. This functionality guarantees that every hour worked is precisely computed and handled for payroll. This indicates that all information from project management tools, communication channels, and payroll systems is combined to function seamlessly together. This removes the necessity for manual data entry and helps in creating a cohesive work atmosphere where all the elements are interconnected effortlessly [6]. The idle time is a vital aspect of the time tracking software that enables monitoring how much time is utilized on tasks and how much time is wasted. This function monitors when an employee is inactive or not engaging with their device or software, providing a clear view of productivity. By monitoring idle time, one can identify inefficiencies and implement solutions to ensure that the team remains productive throughout the workday. With alerts and notifications, one can effortlessly keep track of important events and dates. Having a mobile app is essential for time tracking software. It allows teams to record time from any location, whether in the office, at home, or during a work trip. Mobile apps are beneficial for workers since they can log their hours, modify tasks, and submit timesheets directly from their mobile devices, such as smartphones or tablets. This adaptability allows for the elimination of time loss and guarantees that all individuals are accountable, regardless of where they are located.

2.1.1 Benefits of Time Management and Software Automation

Smart time management employs cutting-edge technologies such as AI, automation, and data analysis to increase how people and organizations utilize their time. The advantages extend past efficiency—improving performance, well-being, and strategic focus. AI systems modify schedules dynamically according to new inputs such as deadlines, emergencies, or changing priorities—facilitating real-time adaptability in rapid environments. Intelligent time management promotes healthier boundaries by automating evening routines, providing reminders to disconnect, or organizing personal priorities. Automated schedules and intelligent planning aid in developing productive habits and sustaining consistency over time [7].

Software automation has emerged as a crucial factor in changing industrial processes, providing a variety of advantages that greatly improve efficiency, productivity, precision, and cost-effectiveness. Through the automation of different processes and tasks, industries can achieve significant enhancements in their overall functions. A key benefit of software automation is enhanced efficiency [8]. Automation simplifies repetitive tasks that were once carried out by hand, enabling faster and more consistent execution. Activities such as data entry, inventory oversight, and production planning, which frequently consume time and are susceptible to human mistakes, can be automated to guarantee that they are performed correctly and on time. Automation minimizes the requirement for manual involvement in these procedures, reducing the likelihood of mistakes and holdups, thus speeding up the entire workflow and allowing sectors to function more efficiently. The decrease in manual involvement also results in a notable reduction in human error, a frequent source of inefficiencies and errors in industrial environments. Automated systems are created to perform tasks with a high level of accuracy, following established guidelines and protocols without variation. This uniformity not only reduces the chance of mistakes but also improves the dependability of the procedures, ensuring that tasks are performed as planned. Consequently, sectors can attain increased levels of productivity and operational efficiency. Enhanced efficiency is an additional major advantage of software automation. Compared with manual methods, automation speeds up operational workflows by executing tasks significantly quicker. For example, automated manufacturing lines can function nonstop, managing large quantities of work without requiring breaks or inactivity. This swift completion of tasks results in a greater output rate and capacity to satisfy rising demand without requiring an extra workforce [9]. Moreover, automation enhances resource allocation by effectively overseeing resources such as labor, materials, and equipment. This optimization helps maximize resource usage, minimize waste, and enhance the overall efficiency of operations. Enhanced precision and quality are also important advantages of software automation. Automated systems feature sophisticated quality assurance methods that assess and track products or processes as they occur. Automated inspection systems utilize sensors and imaging technologies to identify defects and confirm that products adhere to quality standards. This meticulous examination guarantees that only top-quality products are provided to customers, reducing the chances of defects and improving customer satisfaction. The reliability offered by automated systems additionally guarantees that results remain consistent, minimizing discrepancies and upholding elevated quality standards. Cost savings represent a significant benefit of software automation, as they aid in decreasing operational costs and minimizing expenses related to downtime and upkeep. Through task automation, industries can reduce labor expenses linked to manual processes. Automated systems are capable of managing intricate tasks with minimal human involvement, resulting in decreased staffing needs and related expenses [10]. Moreover, automation can reduce downtime by detecting and resolving problems in advance. Predictive maintenance systems utilize data analysis to anticipate equipment failures ahead of time, enabling prompt maintenance and lowering the likelihood of unforeseen breakdowns.

Moreover, software automation increases overall operational effectiveness by uniting different systems and processes into a unified structure. Automation systems can effortlessly integrate with other technologies, including enterprise resource planning (ERP) systems and supply chain management tools, to offer a holistic perspective of operations. This integration fosters enhanced coordination and communication among various departments and functions, resulting in improved decision-making and more efficient processes. In conclusion, software automation provides numerous advantages that greatly enhance industrial processes. Automation increases productivity and speeds up operational workflows by improving efficiency through the simplification of repetitive tasks and minimizing manual input. The increase in precision and quality obtained via automated quality assurance methods guarantees uniform results and decreases the number of defects. Moreover, the savings resulting from decreased operational expenses, minimized downtime, and lower maintenance costs enhance overall financial effectiveness. With the ongoing advancements in technology, the contribution of software automation to industrial success will increasingly expand, equipping industries with essential tools to remain competitive and attain lasting sustainability. Challenges and considerations Software automation has emerged as a crucial element in optimizing industrial processes and productivity, leading to significant enhancements across multiple sectors. The integration and application of automation technologies present considerable challenges and factors that organizations must address to fully gain the advantages of automation. A primary challenge in embracing software automation is system integration. Combining new automation technologies with current systems can be intricate and challenging. Compatibility challenges frequently occur when new software or hardware must work alongside older systems that could have been in use for many years [11]. This challenge is intensified by the requirement to sustain operational continuity throughout the transition phase, which may disrupt current processes. Additionally, merging automation techniques with legacy systems introduces further complications. Legacy systems might utilize outdated technologies that do not work well with modern automation solutions. Enhancing or substituting these systems can be costly and labor intensive, necessitating substantial investments in technology and the workforce. Organizations need to consider the expenses of modernizing outdated systems in relation to the future advantages of automation, frequently resulting in intricate decision-making processes. Moreover, ensuring the operational integrity of legacy systems while adopting new automation technologies requires meticulous coordination and frequently leads to heightened complexity during the transition period [12]. Change management is a crucial part of employee training during the implementation of automation technologies. Automation frequently causes alterations in workflows and job responsibilities, potentially resulting in pushback from employees who are accustomed to conventional practices. Effectively handling this change requires clear communication of automation benefits, addressing employee worries, and offering continuous support. Organizations need to cultivate an environment of flexibility and ongoing education to enable employees to effectively adopt new technologies and work processes. The upfront expenses and investments related to software automation can be significant. Adopting automation technologies frequently necessitates considerable financial investments for acquiring new hardware, software, and associated infrastructure. Calculating the ROI of automation necessitates an in-depth examination of possible cost reductions, enhanced productivity, and better quality. Organizations should thoroughly evaluate their unique requirements and possible advantages to establish if the long-term benefits surpass the initial costs [13]. In addition to financial factors, companies need to consider the possible effects of automation on their operational procedures and employees. Automation may result in substantial alterations to job roles and duties, necessitating modifications in staffing numbers and skill sets. Organizations need to prepare for these shifts and create strategies to facilitate seamless transitions. In summary, although software automation provides significant advantages for enhancing industrial processes and efficiency, it also introduces numerous challenges and factors that organizations need to address. Issues with system integration, training employees, managing change, and initial financial investments are crucial elements that need thorough planning and oversight. By tackling these challenges in a proactive and strategic manner, organizations can harness the complete capabilities of automation technologies to enhance operational excellence and secure lasting success.

2.1.2 Case Studies and Examples

Software automation has transformed industrial processes in numerous industries by increasing efficiency, productivity, and precision [14]. Many case studies showcase effective applications of automation technologies, providing important insights and best practices for improving industrial operations. A notable instance of effective software automation is observed in the automotive sector, particularly within the production processes of general motors (GM). The GM adopted sophisticated automation technologies in its manufacturing processes, employing robotics and automated assembly lines to improve operational effectiveness. The automation of activities such as welding, painting, and assembly enabled the GM to increase production rates while upholding stringent quality standards. Through the incorporation of robotics and automated systems, GM has accomplished notable decreases in cycle times and operational expenses. The automated systems increased worker safety by handling dangerous tasks, which lowered the chances of injuries in the production area. This situation illustrates how automation can result in significant enhancements in efficiency and safety in manufacturing settings. Another significant example can be seen in the pharmaceutical sector, where automation has been essential in optimizing manufacturing procedures. Pfizer, a prominent pharmaceutical firm, adopted automated systems in its production plants to improve the efficiency of drug manufacturing. Automated systems were implemented for activities such as mixing, packaging, and labelling, resulting in faster production rates and enhanced consistency in product quality. The automation of these processes enabled Pfizer to increase production to satisfy rising demand while minimizing the likelihood of human error. This situation emphasizes the ability of automation to increase efficiency and dependability in strictly regulated sectors [15, 16].

In logistics and supply chain management, Amazon serves as a notable example of effective software automation. Amazon's fulfilment centres utilize sophisticated robotics systems to automate the processes of picking, sorting, and packing merchandise. The use of automated guided vehicles (AGVs) and robotic arms has allowed

Amazon to greatly increase the speed of order fulfilment. Automation in Amazon's warehouses has resulted in quicker processing times, increased inventory precision, and lower operational expenses. Through the use of automation, Amazon has revolutionized its supply chain processes, establishing new benchmarks for efficiency and productivity within the logistics sector. The oil and gas sector illustrates the advantages of automation with Shell's implementation of sophisticated monitoring and control systems at its refineries. Shell combined real-time data analytics with automation technologies to enhance refinery operations and increase safety. Automated systems were established for overseeing equipment performance, identifying anomalies, and managing complex processes. This automation allows Shell to increase operational efficiency, minimize downtime, and increase safety by delivering real-time information on equipment status. The effective execution of these systems shows how automation can enhance both operational effectiveness and safety in demanding industrial settings. Insights gained from these successful automation initiatives provide important guidance for other organizations aiming to adopt comparable technologies [17]. A vital takeaway is the significance of detailed planning and integration. Effective automation initiatives necessitate thoughtful evaluation of current systems and processes to guarantee smooth integration. Organizations should evaluate their existing workflows to pinpoint areas where automation can yield the greatest advantages. This includes assessing how well new automation technologies fit with current infrastructure and confirming that they are in line with organizational objectives. A crucial takeaway is the necessity for employee training and managing change. Automation initiatives frequently entail substantial modifications to processes and job responsibilities. Organizations need to allocate resources to training initiatives to prepare employees with the essential skills required to operate and manage new automation systems. Successful change management approaches are essential for addressing employee concerns and promoting a favourable perspective on automation. Ongoing support and effective communication can assist employees in adjusting to new technologies and reduce operational disruptions. Furthermore, organizations should prioritize continuous enhancement and gradual implementation. This enables organizations to recognize and address any problems early and make the required modifications before full implementation. The regular assessment and monitoring of automation systems are vital for ensuring that they consistently fulfil performance requirements and achieve the intended results [18]. Effective strategies for deploying software automation involve performing detailed needs analyses, choosing appropriate technologies, and setting clear goals. Organizations ought to begin by pinpointing particular challenges and sectors where automation can deliver the greatest benefit. This includes evaluating existing procedures, identifying operational difficulties, and establishing quantifiable objectives for automation. Choosing technologies that match these objectives and incorporating them seamlessly into current systems is vital for attaining favourable results. Additionally, companies need to focus on data protection and system dependability in their automation initiatives. Automated systems typically entail gathering and analysing significant volumes of data, rendering data security an essential factor. Enforcing strong security protocols and guaranteeing system dependability are crucial for safeguarding data integrity and ensuring operational continuity. In summary, software automation has shown its ability to enhance industrial processes and efficiency through numerous successful applications in various industries. Insights gained from these projects highlight the significance of detailed planning, workforce training, and ongoing enhancement. By adopting best practices and tackling challenges head-on, organizations can utilize software automation to realize substantial improvements in efficiency and foster innovation within their operations [19].

2.1.3 Future Trends and Innovations

Software automation is advancing swiftly, propelling major progress in industrial processes and efficiency. Emerging trends and advancements in this area are set to revolutionize industry operations, boosting efficiency, cutting expenses, and meeting sustainability objectives. The integration of emerging technologies with Industry 4.0, along with a focus on sustainability and efficiency objectives, is essential in defining the future of software automation. Innovations in automation technologies are poised to transform industrial operations in ways never thought possible. A highly promising advancement is the emergence of sophisticated robotics and artificial intelligence [20]. Robotics powered by AI are becoming increasingly advanced and are able to execute intricate tasks with exceptional accuracy and flexibility. These machines can adapt from their surroundings and gradually increase their efficiency, rendering them essential for intricate and ever-changing industrial functions. For example, collaborative robots, known as cobots, are intended to operate alongside human workers, improving flexibility and efficiency in manufacturing settings. This progress not only enhances productivity but also enables increased customization and flexibility in manufacturing processes. Moreover, the combination of AI and automation systems results in the creation of intelligent factories. AI algorithms can examine extensive data produced by automated systems, facilitating predictive maintenance, real-time optimization, and smart decision-making. Predictive maintenance uses AI to examine equipment data and forecast failures in advance, leading to decreased downtime and lower maintenance expenses [21]. This feature is especially advantageous for sectors with essential machinery that needs continuous oversight. The incorporation of AI into automation systems allows industries to increase operational efficiency, reduce disruptions, and prolong the durability of their machinery. As companies and individuals traverse a more intricate digital environment, the evolution of smart time management and task automation is influenced by cutting-edge technologies, data-informed insights, and user-centered design. Significant trends and advancements include the following.

I. **AI-Powered Personal Assistants**

Next-generation digital assistants such as Siri, Alexa, and ChatGPT are transforming into proactive agents that foresee needs, organize tasks, and streamline decision-making. These tools are becoming more interconnected with calendars, communication platforms, and project management software to provide smooth task execution and prioritization.

II. **Predictive Scheduling and Smart Prioritization**

AI models are trained on user actions, project timelines, and company objectives to forecast the best time slots for tasks [22]. This involves immediate reprioritization in response to changing circumstances—assisting in minimizing procrastination and mental strain.

III. **Automation of Repetitive Tasks**

Robotic process automation (RPA) and workflow automation tools (such as Zapier, Power Automate, and UiPath) are evolving to manage increasingly intricate, multi-step processes. These tools integrate natural language understanding, enabling users to automate tasks through straightforward commands.

IV. **Integration with Wearable and IoT Devices**

Smart time management is moving beyond displays. Wearable devices can now monitor focus, stress, and productivity trends automatically and recommend micro-breaks, task changes, or collaboration periods on the basis of biofeedback [23].

V. **Cross-Platform Synchronization**

Integrated platforms continue to develop, where all tools, calendars, task lists, and messaging applications are linked together. This minimizes the time wasted on switching apps and facilitates a unified dashboard for comprehensive productivity.

VI. **Hyper-Personalization with Machine Learning**

ML algorithms are progressively utilized to develop customized productivity techniques—tailoring approaches such as Pomodoro, deep work, or time blocking according to user history, energy levels, and types of goals [24].

VII. **Voice and Conversational Interfaces**

Automation controlled by voice is becoming increasingly intuitive. Conversational AI is anticipated to assume responsibility for task assignment, meeting recaps, and follow-up planning—entirely through natural language commands.

VIII. **Emotional Intelligence and Sentiment Analysis**

Certain platforms are beginning to incorporate sentiment analysis and emotional AI, allowing them to identify user burnout or disengagement and react with restorative recommendations or task postponements.

IX. **Collaborative Automation in Teams**

Future platforms will automate not only single tasks but also team workflows—allocating roles, issuing reminders, monitoring progress, and recommending task reallocation on the basis of workloads and deadlines [25].

X. **Ethical and Mindful Automation**

As automation increases, there is a heightened focus on ethical design preventing excessive automation, which results in decision fatigue or diminished agency [26]. Upcoming tools emphasize intentional work and support work–life balance rather than just efficiency.

2.2 AI in Boundary Management and Burnout Prevention

Techniques in sentiment analysis and natural language processing powered by AI can identify early indicators of employee emotional strain or burnout. AI systems can identify individuals who may be at risk and provide immediate support, such as access to mental health services or modifications to their work environment. Employee burnout extends beyond a single tough day or even a difficult week at work, and it is a persistent condition that impacts employees worldwide. Burnout is widespread in the current workforce, impacting employees' mental and physical well-being and the prosperity of numerous organizations [27, 28]. Employee burnout cannot be resolved with a restful night or a single enjoyable 2-day weekend. Burnout has evolved from being merely a personal issue to becoming an epidemic affecting organizations globally. In high-pressure sectors such as technology and challenging corporate positions, workers encounter increasing stress, blurred lines between work and life, and fatigue.

Burnout, according to the World Health Organization (WHO), is a condition arising from "ongoing workplace stress that has not been effectively handled." Symptoms may include emotional and physical fatigue, diminished work performance, and an increasing feeling of cynicism. The statistics are concerning. A Gallup survey revealed that 77% of workers indicate that they have faced burnout in their current positions [29].

Burnout affects not only individuals but also team dynamics, extends project timelines, and undermines workplace culture. Numerous organizations are enthusiastic about the prospects of generative AI (Gen AI) enhancing employee productivity and transforming business processes. Nonetheless, they are attempting to balance the possible advantages with the risks, continuing to determine how to reduce the latter. Forrester Research examines how enterprises adopt generative AI (Gen AI) in its report "Generative AI Prompts Productivity, Imagination, and Innovation in the Enterprise." AI-driven systems are capable of assessing video streams from security cameras, employing computer vision to identify and notify unauthorized entries or questionable activities around perimeters such as fences or entrances to buildings [30].

AI can connect with different sensors (motion sensors, infrared beams) to establish an extensive security system, offering immediate alerts and automated reactions. AI can be utilized to establish and oversee boundaries in different settings, such as in project management, where it aids in defining project scopes, distributing resources, and handling risks. Boundary management pertains to how individuals or organizations handle the division or merging of work and personal life, as well as various roles and responsibilities. AI technologies are progressively assisting individuals in smartly navigating, automating, and enhancing these limits.

2.2.1 Key AI Practices in Boundary Management

Supervised learning establishes boundaries via labelled data (e.g., normal versus malicious behavior), whereas unsupervised learning identifies anomalies. When applied in physical boundary management, it processes images or videos to detect boundaries or monitor violations. Analyse written information to create or uphold policy-driven limits. Boundary modifications in changing settings, such as managing traffic flow in smart cities, should be increased [31]. AI identifies and blocks unauthorized access by establishing limits for standard network activities.

Essential AI strategies in boundary management emphasize the use of artificial intelligence to establish, observe, and implement boundaries in areas such as cybersecurity, geospatial systems, organizational processes, and resource management. The fundamental AI strategies designed for boundary management are as follows.

I. **Data-Driven Boundary Definition**

Supervised Learning: Utilize labelled datasets for training AI models to determine boundaries, such as distinguishing normal from anomalous behavior in cybersecurity or delineating property lines in geospatial contexts. Unsupervised Learning: Implement clustering algorithms (e.g., k-means or DBSCAN) to discover natural boundaries within unstructured data, such as customer segmentation or geographic zoning.

II. **Real-Time Monitoring and Anomaly Detection**

Anomaly detection: AI models (e.g., autoencoders, isolation forests) are utilized to oversee activities and identify departures from defined limits, such as spotting intrusions in networks or breaches in restricted areas. Computer Vision: Employ image processing and deep learning (such as convolutional neural networks) to track physical limits via cameras, drones, or satellite images in areas such as border security or city planning. Time series analysis: AI is utilized to evaluate temporal data from sensors or IoT devices to monitor boundary conditions, such as resource consumption limits in energy or network infrastructures [32].

III. **Predictive Boundary Management**

Forecasting Models: Employ interpretable AI models to ensure clarity in the definition and enforcement of boundaries, which is essential for regulatory adherence and trust. Employ interpretable AI models to ensure clarity in the definition and enforcement of boundaries, which is essential for regulatory adherence and trust. Sensor Fusion: Merge information from various sources (e.g., IoT devices, cameras, logs) with AI to generate a consolidated perspective of boundary conditions, enhancing precision in monitoring.

Cross-Domain Analysis: Employing AI to integrate data from different sectors (e.g., combining geographic and financial information) to address complex challenges, such as those found in emergency management or urban planning. Bias Mitigation: Employ AI algorithms that prioritize fairness to ensure that limits (such as in resource allocation or oversight) do not advantage or disadvantage particular groups. Explainability: Utilizing interpretable AI models to guarantee transparency in defining and upholding boundaries is crucial for regulatory compliance and trust [33].

This entails the tactical application of artificial intelligence to assist people and organizations in managing clear boundaries between professional responsibilities, personal life, and various roles. AI-powered tools facilitate intelligent scheduling by automatically reserving focus time, work hours, and breaks, minimizing burnout risk. Virtual assistants and collaboration tools employ contextual intelligence to postpone or silence notifications outside of working hours, aiding in digital disconnection.

Grasping the reasons behind burnout and its diverse symptoms is essential for successfully addressing it. These factors include long-term heavy workloads and sensations of ineffectiveness, and symptoms vary from physical fatigue to behavioral shifts such as isolation. Successful prevention strategies include both personal and organizational methods. Individual self-care routines and stress alleviation are vital, along with organizational initiatives to establish a nurturing workplace, encourage work–life equilibrium, and cultivate an atmosphere of trust and security. In content moderation, it is crucial to address specific stress factors and foster practices that enhance mental health and wellness. Focusing on preventing burnout promotes employee well-being. It improves overall efficiency and the standard of the workplace for companies. Preventing burnout is crucial for fostering a healthy, efficient work environment, especially in high-pressure positions such as content moderation. Acquiring the reasons behind burnout and its diverse symptoms is crucial for effectively addressing it. These factors vary from ongoing heavy workloads to sensations of inadequacy, with symptoms including physical fatigue and behavioral shifts such as isolation [34, 35]. Successful prevention tactics include both personal and organizational methods. Individual self-care routines and stress reduction techniques are vital, similar to organizational initiatives aimed at establishing a nurturing workplace, encouraging work–life harmony, and developing a culture of trust and security. In content moderation, it is crucial to address specific stressors and support methods that enhance mental health and wellness. Focusing on preventing burnout safeguards employee well-being. It improves overall efficiency and the standard of the workplace for companies.

2.2.2 *AI and Prevention of Burnout*

I. **Intelligent Task Management**

AI-powered tools can help workers prioritize tasks, manage workloads, and develop effective schedules. These solutions utilize machine learning techniques to examine work habits, pinpoint obstacles, and deliver tailored suggestions, enabling individuals to optimize their time management and prevent burnout.

II. **Smart Communication and Collaboration**

Chatbots and virtual assistants driven by AI enhance internal communication, alleviating the strain of excessive emails and facilitating immediate collaboration. These smart tools manage standard questions, organize meetings, and promote knowledge exchange, allowing employees to focus on more significant tasks and alleviating stress.

III. **Emotional Well-Being Monitoring**

Techniques for sentiment analysis and natural language processing driven by AI can identify early indicators of emotional distress or burnout in employees. AI systems can recognize people who might be at risk and offer prompt support, including mental health resources or changes to work conditions [36].

IV. **Personalized Learning and Development**

AI-powered learning platforms can provide customized training modules designed to meet the specific requirements of individual employees. By evaluating skill deficiencies and learning styles, these systems offer tailored educational resources and advice and promote career development and job fulfilment, which are essential for preventing burnout (Fig. 2.2).

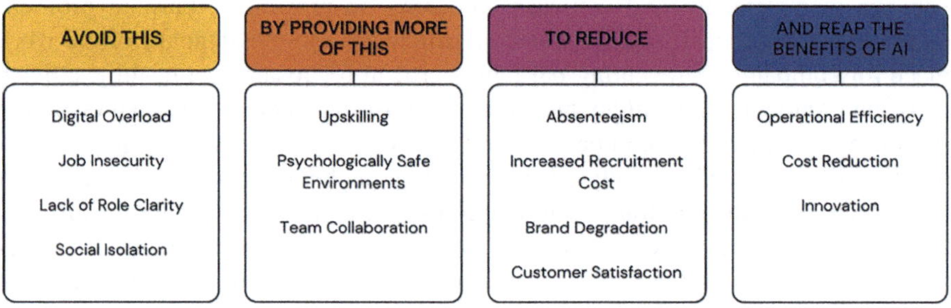

Fig. 2.2 AI in balancing burnout

2.3 AI-Based Personal Assistants for Work–Life Integration

AI-driven personal assistants are virtual software agents intended to help users with different tasks and deliver information via artificial intelligence technologies. These assistants utilize natural language processing (NLP), machine learning, and various AI methods to comprehend and reply to user commands in a human-like way, facilitating task completion with automation and simplicity. Virtual companions frequently utilize sophisticated methods such as natural language generation (NLG) and deep learning, which offer several important characteristics: natural language understanding (NLU) can hold and interpret user requests and instructions in natural language, enabling more seamless interactions. Generative abilities are the active creation of human-like replies, enabling the assistant to deliver contextually appropriate and customized information [37]. Context awareness is the sustaining context throughout various interactions, allowing for more consistent and pertinent replies on the basis of continuing discussions. Ongoing learning through user interactions to enhance performance, grasp preferences, and adjust to changing user requirements. Various tasks are executed according to user demands, ranging from basic commands to intricate, context-driven operations. Conversational dynamics improve dialogue flow through context comprehension, coherence retention, and the elimination of redundant or disconnected replies, leading to a more organic interaction experience.

Enhancing work–life balance can enhance overall well-being, including physical, emotional, and mental health. This includes community, health, and family, as well as career and well-being (Fig. 2.3). A study by the Government Accountability Office (GAO) revealed that long working hours can result in severe health problems, including stroke, coronary heart disease, type 2 diabetes, and heightened alcohol use, although further research is needed to clarify the exact connection between extended work hours and adverse health effects [38].

2.3.1 *Theory and Conceptual Frameworks*

As a result of the changing dynamics between work and life obligations, research in the work–family field has increased significantly over the past twenty years. The amount of research has been driven by heightened global competition leading to mergers, downsizing, and acquisitions; alterations in government policies and regulations; shifts in family socioeconomic demographics; greater complexity in work and family roles; the growth of technology; a rise in dual income/career marriages; a larger number of women joining the workforce; and heightened occupational stress and its effects on employee health and well-being. The foundation for strategically advancing research projects and crafting interventions to address work–life changes should be centred on suitable and well-structured theoretical frameworks. Studies suggest that research and development initiatives in work–family relationships yield

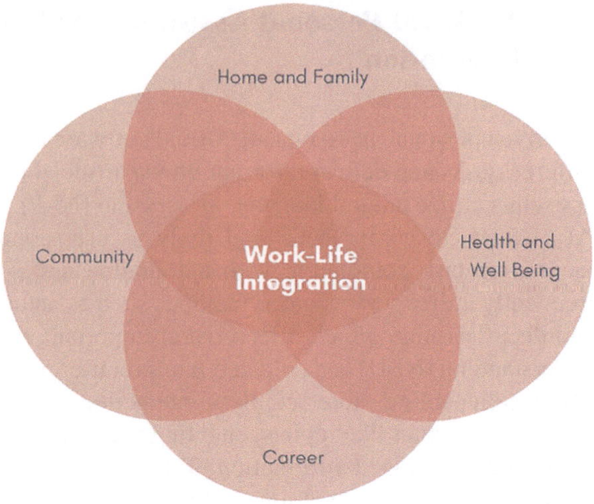

Fig. 2.3 Work–life integration

the best results "if grounded in thoroughly thought-out and well-articulated theoretical and conceptual frameworks." Therefore, we briefly discuss this issue to acquaint the reader of this special issue with commonly utilized work–family theoretical frameworks [39]. Additionally, several of these theoretical perspectives are included by the authors of this special issue, making them pertinent to the audience. The work–family scholarship area includes various theoretical frameworks that researchers and practitioners have utilized to investigate and clarify the connections between work and family life. These theoretical concepts are succinctly described as follows: spillover refers to the overlapping effects (e.g., emotions, values, abilities, actions) between work and family. Spillover experiences may be positive or negative, yet the experiences of work and family are the same—either they are positive or both are negative [40]. Compensation refers to attempts to counteract unfavourable experiences in one area (such as work or family) through greater efforts to pursue positive experiences in other areas (such as family or work). Efforts are carried out via one of two routes. One route involves greater engagement in one area (e.g., work) offset by reduced participation in the alternative area (e.g., family). The alternative route involves choosing the area that provides higher rewards and satisfaction while sacrificing the area that yields minimal benefit. Resource depletion denotes the movement or allocation of a finite set of available resources (e.g., time, energy, focus) from one area (i.e., family) to another (i.e., work), consequently diminishing the accessibility of those resources for use in the domain that initially possessed them. When leftover or unused resources are inadequate, exhausted, or both, it can lead to increased levels of stress, fatigue, and burnout. Enrichment refers to the extent to which experiences, originating from either instrumental pathways (i.e., skills, abilities, values) or affective pathways (i.e., moods), in one area (e.g., work) positively improve the quality of life in another area (e.g., family). Congruence refers to the resemblance between work and family, attributing this similarity to a third variable, such as genetic traits or personality, which influences both areas. Congruence closely resembles spillover, but in congruence, similarity is facilitated by a third variable. On the other hand, spillover represents a direct impact between family and work. Segmentation consists of a boundary that divides the realm of work from family—and family from work [41]. Facilitation involves the degree to which involvement in one area promotes increased participation or activities in a different area. Facilitation comprises abilities, experiences, resources, and knowledge that interact with personal and situational factors, which are transferable and enhance organizational and developmental levels. In these theoretical frameworks, the connections between work constructs and family constructs are frequently viewed by researchers and practitioners as separate and conceptually distinct. This area of research and practice has usually concentrated solely on work and family domains, leading to the exclusion of wider contexts, such as the community, in the study's methodological and intervention designs. Integration embodies a comprehensive approach that includes the effective and efficient alignment of actions and energies among all parties involved who share interests and advantages from employees capable of meeting and transitioning between their personal, work, family, and community responsibilities.

2.3.2 AI Technologies to Increase Work–Life Balance

In the current rapid-paced environment, achieving a healthy balance between work and life is becoming more difficult. Nonetheless, progress in artificial intelligence (AI) is providing fresh avenues to improve this equilibrium by simplifying tasks, enhancing communication, and refining decision-making [42]. Here, how essential AI technologies, including chatbots, natural language processing (NLP), predictive analytics, the automation of routine tasks, and responsiveness to communication requirements, are transforming the workplace to promote improved work–life balance is examined.

I. **Chatbots**

Chatbots have transformed how companies manage customer service requests. These AI-powered assistants function around the clock, delivering immediate answers to questions and addressing problems without the need for human involvement. By managing everyday customer interactions, chatbots save valuable time for employees during work hours, enabling them to concentrate on more intricate tasks that require human insight and creativity. These chatbots also stop employees from being overwhelmed by repetitive inquiries when they are off duty, enabling them to concentrate their time and energy on their personal lives.

II. **Natural Language Processing**

NLP is a field of AI that allows computers to comprehend, interpret, and react to human language in a manner that is both significant and contextually appropriate.

NLP improves communication by offering better transcription services, instant translations, and converting complicated data into clear information [43]. This not only encourages teamwork among employees but can also minimize the troubles and concerns associated with manual transcriptions and delayed translations, thereby promoting a more even workload and reduced stress.

III. **Predictive Analytics**

Predictive analytics employs past data, statistical methods, and machine learning approaches to determine the probability of future results. This technology enables companies to predict user requirements, improve resource distribution, and increase decision-making processes. For example, businesses can anticipate market changes or consu mer desires, allowing for proactive strategies instead of reactive actions. This allows employees to utilize their resources more efficiently, concentrating on projects that are expected to produce the best outcomes, thereby preventing unnecessary resources over time and stress.

IV. **Routine Task Automation**

The automation of repetitive tasks is among the most important ways in which AI contributes to improving work–life balance. By handling repetitive tasks such as data entry, scheduling, and inventory control, AI enables employees to focus their efforts on more significant work. This decrease in repetitive tasks not only increases productivity but also lowers stress levels, resulting in a healthier work atmosphere. Workers are able to have additional leisure time and not stress over work during their time off and are confident that their responsibilities are effectively handled by AI technologies [44, 45].

V. **Adapting to Communication Needs**

The capacity of AI to adjust to personal communication styles is a transformative factor in improving workplace cooperation. Through the examination of communication habits and preferences, AI systems can customize interactions to fit the requirements of individual employees, recommending pertinent contacts and communication methods informed by past exchanges [46]. For example, AI can recommend optimal meeting times according to attendees' availability, prioritize communications, or indicate ideal moments to contact particular people, ensuring that significant messages are always received. These individualized engagements make certain that team members feel acknowledged and appreciated, cultivating a more inclusive and efficient workplace that alleviates stress for everyone concerned.

These diverse AI technologies are reshaping the work–life balance arena and providing substantial advantages by increasing workloads, increasing communication effectiveness, and facilitating strategic choices, all of which lead to greater job satisfaction and better work–life equilibrium. As AI has advanced, it possesses significant potential to transform workplaces further, boosting productivity and employee well-being while creating a future where personal satisfaction and career achievements are harmonized.

2.4 Smart Scheduling and Flexibility Tools

As burnout increasingly becomes commonplace and a relentless work culture jeopardizes well-being, organizations face mounting pressure to reconsider how tasks are accomplished. At Corporate One, it is thought that the future of work should focus not only on performance but also on people. One of the most promising solutions for achieving this balance is AI-driven intelligent scheduling.

AI-driven smart scheduling adds a clever automation and insight component regarding time distribution. It goes beyond simply aligning availability—it understands preferences, predicts overload, safeguards focus time, and encourages flexibility.

I. **Protecting Focus Time**

AI tools can examine calendars and automatically schedule deep work times according to meeting trends and task requirements. This helps employees better manage their mental energy and avoids the frequent context switching that results in burnout.

II. **Preventing Meeting Overload**

AI can identify overly packed days, propose meeting-free periods, and recommend better times for regular check-ins on the basis of team availability and productivity patterns. The outcome? Fewer meetings lead to improved results [47, 48].

III. **Respecting Time Zones and Boundaries**

For global or hybrid teams, intelligent scheduling systems consider time zones and propose meeting times that preserve work–life balance. They are also capable of understanding and honoring personal time limits, instinctively steering clear of meetings that occur late at night or early in the morning.

IV. **Encouraging Breaks and Well-Being**

AI-powered scheduling can remind users to take breaks, encourage teams to maintain healthy work habits, and deter consecutive meetings. As time progresses, this cultivates more wholesome and enhances [49].

2.4.1 Overcoming Common Challenges in AI-Enhanced Scheduling

Although AI scheduling presents significant opportunities for improving work–life balance, companies commonly face various obstacles during both implementation and continual use. Identifying and actively tackling these challenges guarantees that the technology fulfils its potential for better employee experience while maintaining operational requirements. Employees might be skeptical of schedules if they are unclear about the decision-making process of AI, necessitating transparent descriptions of fairness mechanisms.

Effective organizations address these challenges with robust change management strategies, clear communication, and a gradual implementation process that facilitates adaptation and learning. Choosing a solution such as Shyft's AI scheduling assistant, which offers strong support and implementation services, can greatly ease these challenges and speed up the journey to achieve better work–life balance results.

2.4.2 Industry-Specific Work–Life Balance Considerations

Various sectors encounter specific scheduling difficulties that impact work–life balance. AI scheduling solutions need to be set up to address these particular situations while providing significant enhancements to the employee experience. Comprehending these details can help organizations choose and execute the best solution for their employees. Retail must adapt to seasonal changes while allowing employees enough notice to arrange personal commitments. Healthcare needs to balance continuous coverage with proper rest for workers to prevent fatigue and maintain patient safety. By addressing these specialized industry needs, AI scheduling tools provide tailored work–life balance improvements that significantly increase employee contentment. Providers such as Shyft deliver customized shift scheduling solutions that cater to the distinct operational patterns and employee requirements of each industry [50, 51].

2.4.3 Future of AI Scheduling and Work–Life Balance

With the ongoing advancement of AI technology, we can anticipate increasingly advanced scheduling functionalities that improve work–life balance even further. Organizations must remain aware of new trends to preserve their competitive edge in attracting and retaining talent via modern scheduling methods. Numerous advancements are approaching that will transform the employee scheduling environment. Improved predictions of businesses need to reduce last-minute schedule alterations that interfere with personal arrangements. An AI that analyses personal productivity trends to recommend ideal work schedules that sync with biological rhythms.

Top organizations are actively investigating these innovations by collaborating with providers such as Shyft, which consistently improves their AI shift scheduling features. The most innovative companies understand that work–life balance is not merely an employee perk—it is a strategic edge in drawing and keeping high-quality talent in competitive industries.

Entities that have effectively utilized AI scheduling to improve work–life balance adhere to various best practices that optimize employee contentment and operational effectiveness. These methods emphasize careful execution, continuous improvement, and a sincere dedication to employee wellness that goes further than mere scheduling functions. Organizations implementing these strategies foster a positive loop where enhanced work–life balance results in increased employee contentment, lower attrition, and, ultimately, superior customer experiences [52, 53]. By carefully implementing technology in shift management, they convert scheduling from an administrative hassle into a strategic asset for organizational success and employee wellness. AI-driven employee scheduling offers a substantial chance to improve work–life balance while fulfilling operational needs. Through the adoption of smart scheduling tools that emphasize employee choices, enable straightforward shift swaps, and enhance schedule transparency, companies can foster an atmosphere where work support rather than disruption of personal life occurs [54]. The advantages go beyond employee contentment to include better retention, lower absenteeism, and increased productivity, establishing work–life balance improvement via AI scheduling as a strategic business necessity instead of merely an employee benefit.

2.5 Conclusions

AI streamlines repetitive and routine tasks, enabling professionals to concentrate on more creative, strategic, or significant activities. This simplification can lessen stress and allow for a more balanced distribution of time between job responsibilities and personal life. With the help of smart assistants, scheduling applications, and recommendation systems, AI can aid people in organizing their tasks, appointments, and health routines better, customizing recommendations on the basis of individual preferences and requirements. AI-driven platforms enable remote teamwork and interaction, simplifying the ability to work from any location. This adaptability promotes a more balanced work–life dynamic by minimizing travel time and allowing for more versatile work hours. The section emphasized that AI can blur the lines between professional and personal life, particularly when alerts or responsibilities pursue workers beyond conventional office hours. Establishing healthy digital boundaries is essential, ensuring that AI acts as a means for balance rather than excessive labor. AI ought to enhance, not substitute, human decision-making. The chapter asserts that careful application and ethical factors are crucial to maintain the beneficial influence of AI on well-being, serving as a significant facilitator for attaining work–life balance; however, its success relies on intentional utilization, defined limits, and continuous assessment. Organizations and individuals need to work together to ensure that AI promotes productivity, adaptability, and personal well-being instead of leading to burnout or imbalance.

References

1. Gao, K., & Zamanpour, A. (2024). How can AI-integrated applications affect the financial engineers' psychological safety and work–life balance: Chinese and Iranian financial engineers and administrators' perspectives. *BMC Psychology, 12*(1), 555.
2. George, A. S. (2024). Automated futures: Examining the promise and peril of AI on jobs, productivity, and work–life balance. *Partners Universal Innovative Research Publication, 2*(6), 1–17.
3. Tariq, M. U. (2025). AI and work–life balance: Transforming employee wellbeing in the modern workplace. In *Strengthening human relations in organizations with AI* (pp. 85–110). IGI Global Scientific Publishing.
4. Korrapati, R. (2024). AI-driven enhancement of employee experience: A strategic integration within human resources.
5. Oladele, O. K. (2023). Data-driven work culture: How AI enhances employee well-being and reduces occupational stress.
6. Turčinović, M., Vujko, A., & Mirčetić, V. (2025). Algorithmic management in hospitality: Examining hotel employees' attitudes and work–life balance under AI-driven HR systems.
7. Azad, M., & Malhotra, C. (2020) Work–life balance driven by AI: The next big thing in employee monitoring. *Foundations of, 261.*
8. Heikkilä, H., & Okkonen, J. (2021). AI driven competency development at the threshold of working life. In *Proceedings of the 2021 Nordic learning analytics (summer) institute, NLASI 2021.* CEUR-WS.
9. Agrawat, P., & Deshani, S. H. (2025). Future of work and well-being in the AI-driven service sector: A qualitative study. In *Practices, challenges, and deterrents in workplace wellbeing: Strategies for building resilient and thriving workplaces* (pp. 67–92). IGI Global Scientific Publishing.
10. Rao, T. V. N., Patel, G. S., Bhavana, S., & Swetha, C. (2025). AI-driven decision support system for workplace wellbeing. In *Insights on wellbeing and happiness in the workplace* (pp. 1–18). IGI Global Scientific Publishing.
11. Rydén, P., & El Sawy, O. (2022). Real-time management: When AI goes fast and flow. In *Platforms and artificial intelligence: The next generation of competences* (pp. 225–243). Springer.
12. Saardloun, P., Mahadumrongkul, N., Surajarus, K., Suppataratarn, K., Horaraung, K., Wonggulya, N., et al. (2024). Designing a user-centric AI-driven mobile app for personalized time management: Integrating machine learning and design thinking. In *Proceedings of the 2024 23rd international symposium on communications and information technologies (ISCIT)* (pp. 212–217). IEEE.
13. Gyory, J. T., Soria Zurita, N. F., Martin, J., Balon, C., McComb, C., Kotovsky, K., & Cagan, J. (2022). Human versus artificial intelligence: A data-driven approach to real-time process management during complex engineering design. *Journal of Mechanical Design, 144*(2), 021405.
14. Ronak, B. (2024). AI-driven project management revolutionizing workflow optimization and decision-making. *International Journal of Trend in Scientific Research and Development, 8*(6), 325–338.
15. Schrettenbrunnner, M. B. (2020). Artificial-intelligence-driven management. *IEEE Engineering Management Review, 48*(2), 15–19.
16. Haghsheno, S. (2021). AI-driven project management in software engineering. *International Journal of Scientific Development and Research, 6*(1), 299–308.
17. Barbar, C., Bass, P. D., Barbar, R., Bader, J., & Wondercheck, B. (2022). Artificial intelligence-driven automation is how we achieve the next level of efficiency in meat processing. *Animal Frontiers, 12*(2), 56–63.
18. Sodiya, E. O., Umoga, U. J., Amoo, O. O., & Atadoga, A. (2024). AI-driven warehouse automation: A comprehensive review of systems. *GSC Advanced Research and Reviews, 18*(2), 272–282.

19. King, R., & Zenil, H. (2023). A framework for evaluating the AI-driven automation of science. *Artificial Intelligence in Science, 113*, 331.
20. Natarajan, D. R. (2020). AI-generated test automation for autonomous software verification: Enhancing quality assurance through AI-driven testing. *International Journal of HRM and Organizational Behavior, 8*(4), 89–103.
21. Ikwuo, A. K., Nworie, G. O., & Moedu, V. O. (2024). Implementation of AI-driven automation: A game-changer in accounting research. *International Journal of Financial, Accounting, and Management, 6*(3), 385–398.
22. Paulose, R., & Neelanath, V. (2024, May). Generative AI-driven automation of business process reimagination. In *Proceedings of the 2024 IEEE recent advances in intelligent computational systems (RAICS)* (pp. 1–6). IEEE.
23. Adamopoulos, I., Valamontes, A., Tsirkas, P., & Dounias, G. (2025). Predicting workplace hazard, stress and burnout among public health inspectors: An AI-driven analysis in the context of climate change. *European Journal of Investigation in Health, Psychology and Education, 15*(5), 65.
24. Zhang, T., & Liu, X. (2025). Tracking the evolving impact of AI-driven learning platforms on EFL students' burnout, emotional challenges, and well-being: A longitudinal growth curve analysis. *Innovation in Language Learning and Teaching, 53*, 1–21.
25. Liu, H., Ding, N., Li, X., Chen, Y., Sun, H., Huang, Y., et al. (2024). Artificial intelligence and radiologist burnout. *JAMA Network Open, 7*(11), e2448714–e2448714.
26. Majumder, S., & Dey, N. (2022). *AI-empowered knowledge management*. Springer.
27. Meduri, K., Nadella, G. S., Gonaygunta, H., Kumar, D., Addula, S. R., Satish, S., & Rehman, S. U. (2024). Human-centered AI for personalized workload management: A multimodal approach to preventing employee burnout. *Journal of Infrastructure, Policy and Development, 8*(9), 6918.
28. Shankar, R. K. (2019). Reducing IT workforce burnout through AI-driven workload monitoring and personalized stress management solutions.
29. Harris, E. (2024). AI-drafted responses to patients reduced clinician burnout. *JAMA, 331*(17), 1440–1440.
30. Baek, G., & Cha, C. (2025). AI-assisted tailored intervention for nurse burnout: A three-group randomized controlled trial. *Worldviews on Evidence-Based Nursing, 22*(1), e70003.
31. Enzmann, D. R. (2024). Physician burnout: A hidden cause. *Academic Radiology, 31*(2), 718–723.
32. Barnabas, B., & Marvelous, O. (2025). Transforming IT burnout prevention with centralized work pattern monitoring powered by AI.
33. Emmanuel, M., & Liang, W. (2020). AI-driven workload optimization using centralized monitoring to mitigate IT industry burnout with MongoDB atlas.
34. Majumder, S., & Dey, N. (2024). *The vogue of managing people in workplace* (pp. 1–134). Springer.
35. Sajidha, S. A., Vijayanand, D., Rajagopal, M., Sambasivam, D., & Nisha, V. M. (2025). Decoding work–life balance conundrum: A meta-learning prediction approach. In *Creating AI synergy through business technology transformation* (pp. 313–336). IGI Global.
36. Senarathne-Tennakoon, K. U. (2021). Empowerment or enslavement: The impact of technology-driven work intrusions on work–life balance. *Canadian Journal of Administrative Sciences/Revue Canadienne des Sciences de l'Administration, 38*(4), 414–429.
37. Das, S., Chakraborty, S., Sajjan, G., Majumder, S., Dey, N., & Tavares, J. M. R. (2022). Explainable AI for predictive analytics on employee attrition. In *International conference on soft computing and its engineering applications* (pp. 147–157). Springer.
38. Teya, U. B., & Chowdhury, M. (2023). A machine learning-based intelligent system for women's work–life balance prediction and a smart mobile application for women's daily life assistance. *International Journal of Smart Technology and Learning, 3*(3–4), 325–354.
39. Karwa, K. (2025). The modern dilemma: Navigating work life balance. *Journal of Psychology and Behavior Studies, 5*(1), 15–26.

40. Kumari, A., Kanwal, P., Singh, S., Patel, M., Kaur, G., Mittal, S., & Randhawa, S. (2025). Remote work realities and public sector burnout: Addressing mental health, work–life balance, and sustainable solutions. In *Global work arrangements and outsourcing in the age of AI* (pp. 47–72). IGI Global Scientific Publishing.
41. De Cieri, H., Holmes, B., Abbott, J., & Pettit, T. (2005). Achievements and challenges for work/life balance strategies in Australian organizations. *The International Journal of Human Resource Management, 16*(1), 90–103.
42. Yockelson, S. R., Bruns, K., Hemati, K., & Leong, R. (2024). Assessing work/life balance for ACCM trainees. *ASA Monitor, 88*(4), e6–e7.
43. Deep, P. D., Ghosh, N., & Chen, Y. (2025). Faculty burnout in higher education: Effects on student engagement, learning outcomes, and artificial intelligence-driven institutional responses. *Journal of Educational and Developmental Psychology, 15*(1), 11542.
44. Saif, A., Rehman, T., & Alam, S. (2020). Contours of emerging technology on work life balance and economic well-being. *GMJACS, 10*(1), 14–14.
45. Majumder, S., Chowdhury, S., Dey, N., & Santosh, K. C. (2022). Balance your work–life: Personal interactive web-interface.
46. Majumder, S., Ghosh, S., & Polkowski, Z. (2023). A brief introduction to human resource management. In *Artificial intelligence techniques in human resource management* (pp. 1–25). Apple Academic Press.
47. Pavuluri, S., Sangal, R., Sather, J., & Taylor, R. A. (2024). Balancing act: The complex role of artificial intelligence in addressing burnout and healthcare workforce dynamics. *BMJ Health and Care Informatics, 31*(1), e101120.
48. Gupta, D., Bhanumathi, P., Hemavathi, G. P., & Jayashree, S. (2024). Role of technology in enhancing the work–life balance of lecturers in management institutions. In *Business sustainability with artificial intelligence (AI): Challenges and opportunities* (pp. 15–25). Springer.
49. Anastasiia, M. (2021). Work–life balance expectations of young talents: Projections to the year 2026.
50. Warren, T. (2021). Work–life balance and gig work: 'Where are we now' and 'where to next' with the work–life balance agenda? *Journal of Industrial Relations, 63*(4), 522–545.
51. Vyas, L. (2022). "New normal" at work in a post-COVID world: Work–life balance and labor markets. *Policy and Society, 41*(1), 155–167.
52. Murthy, M. R., & Gernal, L. (2024). Work–life balance initiatives to promote employee performance: A case for an organization in UAE. In *Achieving sustainable business through AI, technology education and computer science: Volume 1: Computer science, business sustainability, and competitive advantage* (pp. 289–298). Springer.
53. Radha, K., & Rohith, M. (2021). An experimental analysis of work–life balance among the employees using machine learning classifiers. arXiv preprint arXiv:2105.07837.
54. Parthasarathy, K., Ayyadurai, R., Panga, N. K. R., Bolla, R. L., & Ogundokun, R. O. (2025). Enhancing the emotional intelligence for business and work–life balance using fuzzy logic adaptive model of emotions. *Service Oriented Computing and Applications, 14*, 1–13.

Chapter 3
AI Technologies Driving Wellness Transformation

Predictive analytics influences all areas of healthcare and creates thrilling opportunities, ranging from timely interventions to more personalized, considerate, and interactive patient experiences. As a potent instrument for improving healthcare services, it provides insights that are crucial for fostering efficiency, precision, and better patient results. The prospective path of predictive analytics in healthcare is equally encouraging and demanding. Innovative technologies such as AI greatly enhance predictive analytics functions, and healthcare organizations need to adopt this technology to remain leaders in healthcare advancement. Achieving success in predictive analytics necessitates a solid foundation, and an advanced architecture (e.g. lake house) is the initial step in utilizing this technology effectively. The incorporation of these technologies into healthcare processes is essential for maximizing their benefits and enhancing patient outcomes and operational effectiveness. By combining thoughtful considerations with innovative ideas, advocates of technology and healthcare trailblazers can work together to create a more predictive and proactive healthcare system [1, 2].

3.1 Machine Learning for Personalized Wellness Plans

The incorporation of AI technologies into wellness systems signifies a crucial change in the ways in which people, organizations, and communities view health and well-being. Through predictive analytics that foresee health risks to tailor suggestions that enhance lifestyle decisions, AI enables individuals to actively manage their physical, mental, and emotional well-being. The utilization of information from various sources—such as wearable technology, health records, genetic data, and environmental influences—AI facilitates a comprehensive and integrated method for health management.

S. Majumder and N. Dey, *AI-Driven Wellbeing to Enhance Lives and Work Environments*, Studies in Computational Intelligence 1257,
https://doi.org/10.1007/978-3-032-15735-5_3

Nonetheless, this change has gone further than just the technology alone. The true strength of AI is its capacity to promote a more human-focused wellness culture—prioritizing prevention instead of treatment, personalization rather than standardization, and accessibility in place of exclusivity. AI-powered wellness solutions can increase productivity, lower healthcare expenses, and foster a more engaged and resilient workforce for organizations. However, along with this promise comes accountability. The design of ethical AI, data protection, openness, and fair access should remain at the forefront to guarantee that these improvements benefit everyone without worsening current disparities. As AI advances, its function in wellness progressively transitions from a supportive resource to a fundamental ally in promoting healthier lives and communities [3]. Ultimately, the change led by AI is not just about technology redefining wellness—it is about redefining wellness itself. By merging human compassion with artificial intelligence, we can create a future where well-being is not an exclusive benefit but rather a reality accessible to everyone.

Figure 3.1 depicts the various origins of healthcare data and their role in creating a holistic health information ecosystem. At the core is healthcare data, which are gathered from various sources depicted around the circle. Genomic data refer to information derived from genetic testing, DNA sequencing, and genomic studies. Medical images consist of data obtained from diagnostic imaging techniques such as X-rays, MRIs, and CT scans [4]. Medical records consisted of patient health histories, prescriptions, laboratory results, and hospital documentation. Environmental data refer to details regarding environmental elements such as air quality, water cleanliness, and climate, which can affect health. Internet data offer health-related insights from digital platforms, telemedicine services, and search behaviors. Social media data encompass insights gained from patient interactions, health-related conversations, and online shared experiences. Wearable devices display information from fitness trackers, smartwatches, and various health monitoring tools (heart rate, steps, and sleep). Mobile phone data refer to health applications, location tracking, monitoring of activities, and interactions with healthcare providers [5].

3.1.1 Role of Artificial Intelligence in Wellness and Healthcare

AI is becoming a more important factor in improving overall well-being and particularly influencing workplace wellness through diverse innovative tools and systems. The utilization of AI technologies enables the development of more customized, scalable, and accessible solutions for health and wellness. Similarly, AI has certain trade-offs and dangers. There are several areas where AI enhances well-being in both everyday and work environments: (i) General Wellbeing: Personal Health Monitoring: Devices like Oura, Fitbit, Apple Watch, and Whoop utilize AI to track a person's health data in real time, including heart rate, sleep habits, and activity levels, delivering tailored feedback and suggestions for healthier living; and (ii) Mental

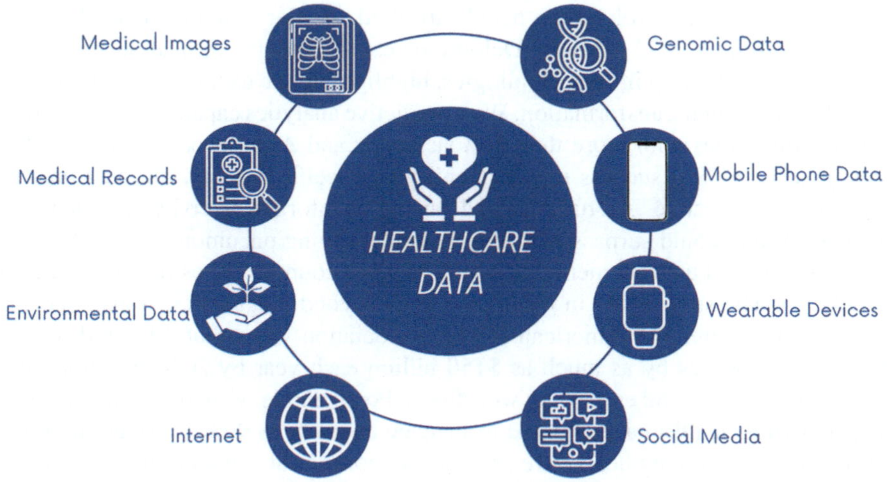

Fig. 3.1 Machine learning for predictive analytics in healthcare

Health Support: Wysa, Woebot, Replika, and Youper serve as AI chatbots and virtual assistants that provide psychological support and counselling, increasing access to mental health resources. They offer prompt, stigma-free assistance for anxiety, depression, and stress, directing users to suitable resources when needed. (iii) Nutrition and diet planning: AI tools like Lose It, PH360, Yazio, MyMealPlanAI, Mealime, and Noom Coach can examine nutritional requirements and propose tailored meal plans, aiding individuals in reaching their health and wellness objectives. This can be especially useful for controlling conditions such as diabetes or for managing weight. Employee Experience Platforms: Tools like AFFINITY OS, Peakon, Culture Amp, Limeade, and Glint leverage AI to assess employee feedback instantly, offering insights into engagement and satisfaction levels. This helps businesses address problems quickly and enhances the overall work atmosphere. While AFFINITY OS merges AI with a human interface, numerous AI platforms offer complete AI interfaces. (vi) Stress Management and Mindfulness: AI-driven applications can deliver tailored mindfulness and stress management activities, assisting team members in lowering stress, enhancing focus, and increasing productivity. These can be effortlessly incorporated into everyday work practices. AI-driven applications include HeadSpace, Calm, Wysa, Youper, and Aura. (v) Ergonomics and physical well-being: AI can evaluate workplace ergonomics and recommend enhancements to minimize physical stress and avert injuries. For example, cameras powered by AI (e.g. knowella), can evaluate posture and suggest modifications instantly [6–8].

There exists a vast range of applications for AI to customize health and wellness initiatives aimed to promote workplace well-being. There are possible drawbacks and risks linked to its application in these fields. Considering these concerns is crucial when thinking about implementing AI in the workplace.

Recently, AI has revolved the healthcare field, similar to a superhero debuting in a major film. A 2022 study by Deloitte revealed that 71% of healthcare leaders were actively investing in AI technologies, highlighting the excitement and urgency linked to this digital transformation. With predictive analytics capable of anticipating patient admissions with more than 90% accuracy and AI-assisted diagnostics that have reached a 99% success rate in identifying specific cancers, the data present a persuasive narrative. A study carried out by Stanford University revealed that their AI system could surpass radiologists in identifying pneumonia from chest X-rays, a significant advancement that could preserve countless lives if it is embraced widely [9]. The story of AI in healthcare goes beyond diagnostics; it also encompasses patient care. The American Health Association states that AI could reduce healthcare expenses by as much as $150 billion each year by 2026 by improving operational choices and enhancing workflows. For example, virtual health assistants (e.g. Babylon Health) have claimed to manage as many as 99% of triage inquiries autonomously, enabling healthcare providers to concentrate on more intricate cases. Additionally, the incorporation of AI into telehealth services has resulted in an impressive 80% rise in patient engagement, demonstrating that the combination of technology and care can significantly increase the patient experience. The revolution persists, as AI leads us into a future where healthcare is not merely reactive but also anticipatory and tailored, providing patients with the care they require even before they realize they need it [10, 11].

3.1.2 Enhancing Patient Engagement Through AI Technologies

In a world where technology is swiftly advancing, healthcare providers are utilizing the capabilities of artificial intelligence (AI) to enhance patient engagement such as never before. An eye-catching figure shows that ~70% of healthcare organizations are now investing in AI technologies, expecting market growth from $4.9 billion in 2020 to an impressive $45.2 billion by 2026. A patient named Sarah, who previously struggled to manage her medications because of her busy lifestyle, was encouraged [12, 13]. Thanks to AI-powered applications, Sarah now receives customized reminders and health advice suited to her unique requirements, greatly enhancing her compliance with treatment. Research indicates that when patients engage in their treatment with these innovative tools, their satisfaction levels rise by as much as 30%. Furthermore, AI technologies improve patients' experiences and offer healthcare providers important insights. A new study by the Mayo Clinic revealed that hospitals utilizing AI analytics experienced a 25% decrease in readmission rates because of their ability to detect potential health decline early [14]. In this situation, John, a patient with a chronic illness, is continuously monitored via wearable technology powered by AI. The system continually monitors John's health data, notifying his healthcare team if his condition starts to vary. This proactive method not only empowers patients (e.g.

John) but also enables clinicians to concentrate their efforts where they are most necessary, fostering a more effective and patient-focused healthcare system.

3.1.3 Ethical Considerations in AI-Driven Health Solutions

In the swiftly changing realm of AI-powered health solutions, ethical issues arise as crucial central concerns. A recent study conducted by the Pew Research Center revealed that 48% of American adults voiced concerns regarding the possibility of biased algorithms in healthcare, resulting in unequal treatment. For example, in 2019, a popular AI system for examining medical images was shown to have lower accuracy for specific demographics, emphasizing the urgent requirement for inclusivity in datasets [15]. The tale of a young woman named Sarah, whose ailment remained undetected because of algorithmic bias, highlights this urgency. These shortcomings not only threaten patient outcomes but also diminish public confidence in AI technologies, with an astonishing 60% of survey participants stating that ethical oversight is essential in the creation of these systems.

Additionally, the financial implications are significant, as the worldwide AI healthcare market is anticipated to hit $188 billion by 2030, on the basis of Fortune Business Insights. However, as organizations strive to innovate, a troubling contradiction appears: although 67% of healthcare leaders acknowledge the need for ethical structures for AI incorporation, only 22% have established official protocols. The tragic story of a startup that favoured profits over ethical leadership and eventually encountered a class-action lawsuit demonstrates this danger [16, 17].

3.1.4 Future Trends in AI and Machine Learning for Personalized Health

As the sun lowers beyond the vibrant pathways of healthcare advancement, a novel age emerges fueled by artificial intelligence (AI) and machine learning (ML). A future in which healthcare professionals possess not only medical history but also current data regarding lifestyle, genetics, and the environment should be envisioned, allowing them to customize treatments uniquely. A new report by Gartner forecasts that by 2025, 75% of healthcare institutions will adopt AI-driven technologies for individualized patient care, greatly increasing treatment results. Moreover, research conducted by McKinsey and Company indicated that AI-driven personalized care has the potential to save the U.S. healthcare system as much as $150 billion each year by reshaping the prevention and management of chronic diseases.

However, the fabric of personal health shaped by AI and ML is not just conceptual; it is interlaced with remarkable data that underscore its possible effects. Research suggests that AI applications in healthcare might generate more than $150 billion in yearly savings for the U.S. health economy by 2026, increasing expenses. A survey from the MIT Sloan Management Review indicated that 83% of leaders in healthcare believe that AI significantly enhances the quality of patient care, promoting both efficiency and a greater comprehension of each patient's specific needs [18]. At this crossroads of technology and health, the movement towards more tailored, data-informed care is not merely hopeful—it is actively happening right now. Additionally, ethical considerations and data privacy issues related to AI and machine learning need to be addressed to guarantee the responsible use of these technologies. Clarity in algorithms, compliance with data privacy laws, and building trust between users and tech providers are essential for the effectiveness of personalized health programs. While it traverses this inventive terrain, cooperation among technologists, healthcare experts, and policymakers will be essential to build a framework that enhances the advantages of AI while protecting personal rights. In the end, the combination of technology and personalized health care has the potential to not only change personal health journeys but also influence the future of worldwide health.

3.1.5 Case Studies of Successful AI Integration in Wellness Programs

In the vibrant city of San Francisco, a technology startup called WellTech was working to improve employee wellness by leveraging innovative AI integration. Through the use of machine learning algorithms, they created a tailored wellness app that examines employees' habits, choices, and health data. Since its inception, WellTech has noted an impressive 30% decrease in absenteeism and a 25% increase in overall employee satisfaction. A survey by the Global Wellness Institute revealed that businesses investing in workplace wellness initiatives gain a $4.00 return for each dollar invested, highlighting how AI can promote a healthier workforce while also increasing productivity and morale [19, 20].

At the center of New York City, a financial company called FinCorp adopted AI-powered analytics to develop a more adaptive wellness program. By gathering information on stress levels, work hours, and engagement via wearable devices, FinCorp designed interventions that aligned with their employees' needs. Following the implementation of these changes, the company noted a significant 40% increase in engagement with wellness programs and a notable 15% reduction in health-care expenses. A report by the Employee Benefit Research Institute revealed that 67% of employers are currently incorporating AI technologies into their wellness programs, suggesting that the future of corporate health efforts is undeniably linked to technological progress [21, 22].

The incorporation of artificial intelligence and machine learning into personalized health and wellness frameworks signifies a major advancement in how people handle personal health management. These technologies facilitate the examination of extensive personal health data, permitting a more customized approach to wellness that takes into account the distinct needs and preferences of individuals. By utilizing predictive analytics and real-time monitoring, AI-powered systems are improving preventative care and enabling individuals to take control of their health journeys. With the ongoing evolution of these technologies, people can anticipate enhanced health results and a more preventative stance on wellness.

3.2 Natural Language Processing for Mental Health Support

NLP can potentially lead to significant changes in mental health. NLP methods give rise to approaches for examining text that could lead to important mental health results, both personally and collectively. It can aid in predicting and addressing consequences linked to any form of decline in mental health metrics, primarily depression and suicide, resulting in improved social outcomes. It also creates innovative methods for preventing, identifying, diagnosing, and managing different mental health issues. This article offers a comprehensive examination of how NLP methods are applied in mental health [23]. It will concentrate on both established techniques and new developments in this field. This study seeks to investigate the diverse applications of NLP in mental health, including but not limited to sentiment analysis, emotion detection, and predictive modelling. It examines the usefulness of these applications in screening, diagnosing, treating, and monitoring mental health. The piece explores how NLP aids in recognizing particular mental health issues like depression, anxiety, PTSD, and schizophrenia, along with its incorporation into therapeutic environments and self-help resources, like online chatbots that provide support for behavioral and mental challenges.

The significant application of NLP in mental health is a plethora of applications of NLP that can be considered in the field of mental health. A summary of some of these applications is provided here. Potential applications of sentiment analysis, a practice in NLP, include the detection of mood fluctuations and mental states by analysing user-generated content. Emotion recognition in text involves identifying specific feelings to assist in affect assessments [24]. Depression and suicide risk detection, assessment, and prevention, including the detection of suicidal ideation in text data, are also prospective applications. The text data fundamentally include social media posts and online communication. CBT chatbots automate therapeutic interventions by acting as digital or virtual therapists. Automated analysis of therapy transcripts helps us discern corrective insights.

NLP is used in mental health evaluations via surveys by linguistically analysing text-based replies to online content and publications. Recognizing PTSD symptoms entails noticing indicators of PTSD in written text, particularly in data produced by veterans, as they represent a group most vulnerable to PTSD. Predictive modelling for mental health emergencies employs NLP to foresee potential crises, often including suicide, offenses such as manslaughter, situations involving self-harm or harm to others, and malicious damage to property through arson and other methods. Substance abuse detection involves identifying talks concerning the sale, purchase, usage, and addiction of drugs in online forums; phone calls; and conversations accidentally overheard and recorded in public areas. Tracking the risk of relapse involves overseeing recovery and the potential for relapse in patients. In various applications of NLP within mental healthcare, the early identification of psychological disorders [e.g. bipolar disorder (BPD)] is improved by NLP methods that have the ability to recognize the presence of these conditions through text and speech. NLP consists of various techniques for examining applications in mental health. Sentiment analysis identifies emotional nuances in written content. This aids in several areas related to mental wellness, particularly the recognition of indicators of anxiety. Subject modelling clarifies prevalent themes in conversations about mental health. Text classification organizes text data and aids in uncovering mental health insights from it [25]. The data consists of social media updates; records from hospitals, such as patient notes and therapy transcripts; suicide notes; and voice messages from individuals just prior to their suicide attempts, among other sources. Named entity recognition increases and improves the volume and significance of information extracted from clinical records.

Chatbots that utilize sophisticated language models, particularly large language models (LLMs), (e.g. Mistral), replicate therapeutic dialogues. They are utilized in automating medical diagnostics and offering self-help guidance to caregivers of patients receiving mental health therapies. Speech and audio analyses identify emotional conditions by observing shifts in tone and pitch. Longitudinal analysis observes language use across time to assess the development of mental health and the efficacy of interventions, whereas cross-sectional analysis examines the similarities and differences in mental health results across different regions, populations, languages, cultures, ethnicities, and other factors of social identity [26].

The incidence of mental health disorders is increasing, impacting millions of people globally. At present, a majority of people who need mental health care do not receive any treatment, with even more restrictions on access for ethnic minorities, low-income groups, and those living in rural areas. In numerous instances, mental health problems can be successfully addressed or, at times, avoided through prompt identification and intervention. Nevertheless, numerous individuals do not obtain the assistance they need until their mental health issues have intensified. Mental health challenges, including depression, anxiety, posttraumatic stress disorder (PTSD), and other psychiatric conditions, have emerged as a worldwide crisis, impacting millions of people from diverse demographics. The World Health Organization (WHO) states that depression is presently the top cause of disability globally, affecting more than 280 million individuals [27]. Moreover, anxiety disorders, encompassing generalized anxiety disorder (GAD), panic disorders, and phobias, affect ~264 million individuals worldwide. PTSD, which is commonly linked to traumatic events such as war, accidents, and abuse, impacts ~3.5% of U.S. adults annually, although its occurrence can be significantly greater in particular groups like veterans. These shocking figures highlight the importance of prompt intervention and proficient management of mental health conditions. Early identification of mental health problems is crucial for enhancing outcomes. Research indicates that prompt intervention can lessen the duration and intensity of symptoms, avert relapse, and improve the overall quality of life of patients. Regrettably, conventional diagnostic techniques depend significantly on self-report questionnaires and face-to-face interviews, both of which are vulnerable to bias, inaccuracies, and limited accessibility. There is a pressing demand for additional innovative, accessible, and objective diagnostic methods to enhance current strategies [28]. Natural language processing (NLP) offers a promising opportunity for addressing these shortcomings. NLP, a field of artificial intelligence (AI), centers on the relationship between computers and human languages, allowing machines to comprehend, analyse, and react to text and speech produced by humans. In recent years, NLP has been utilized across multiple fields, such as healthcare, to examine unstructured text data and extract insights into patients' health. In mental health, NLP can be utilized to identify nuanced patterns in language that reflect psychological conditions, such as negative emotions, incoherent speech, or cognitive biases [29]. For example, people experiencing depression might show distinct language habits, such as increased use of negative emotional terms and a preference for first-person pronouns, reflecting a self-centered perspective. NLP is a field of artificial intelligence that allows computers to comprehend, analyse, and produce human language. In the realm of mental health assistance, NLP is employed to examine and understand extensive amounts of text data—such as therapy session transcripts, social media content, or patient diaries—to extract insights regarding a person's mental health condition and to aid interventions.

Certain systems leverage NLP to both detect mental health issues and produce constructive feedback or interventions, alleviating the workload of human clinicians and offering prompt assistance to those requiring help. Conversation analysis: NLP tools can examine the content and format of discussions in therapy or support groups, assisting clinicians in grasping patient development and customizing interventions more successfully [28]. By utilizing big data from platforms such as social media or electronic health records, NLP can enhance the screening and diagnosis of mental health issues, increasing the accessibility and scalability of mental health care.

NLP holds great potential for mental health assistance but encounters obstacles such as maintaining privacy, addressing the intricacies of human emotions, and preventing biases in language interpretation. The use of AI-powered language analysis for mental health support includes detecting, monitoring, and intervening in mental health concerns and providing scalable, personalized, and prompt support for both individuals and healthcare providers [30].

Figure 3.2 illustrates a crisis detection and alert system for patient communications, which uses a mix of automated filters, machine learning algorithms, and human oversight. Every Patient Message—Every message sent by patients is recorded in the system. Primary Filter—crisis keywords filter. The system reviews all communications for specific crisis-related terms or phrases (e.g., "suicide," "emergency," "assistance urgently"). Only messages including these keywords proceed. Time Filter—Recent Communications. Messages are additionally filtered to capture only those dispatched in the last 10 min, guaranteeing immediate responsiveness.

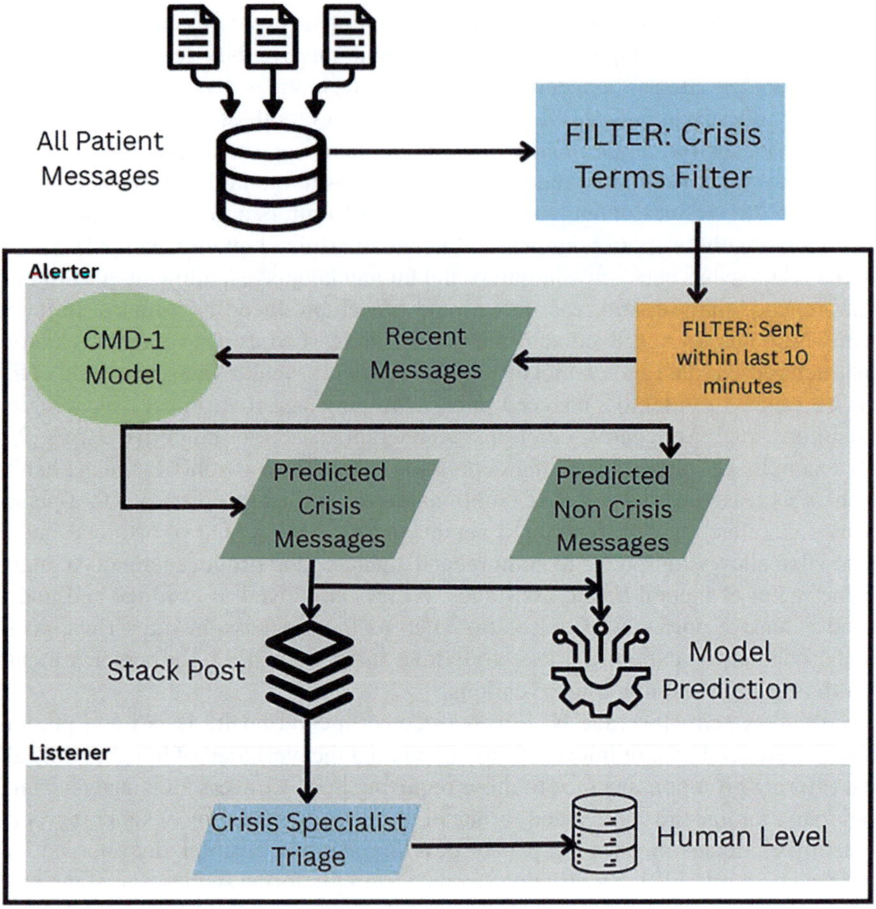

Fig. 3.2 Using natural language processing to detect mental health crises

Alerter Module-In the "Alerter" area: Model outcomes. Forecasted crisis notifications → dispatched as Slack messages to instantly notify appropriate teams. Crisis and noncrisis forecasts are recorded in Model Predictions for monitoring and assessment. Listener module—At this point, crisis specialists assess (triage) the highlighted messages.

3.3 Wearable Devices and AI Integration

Wearable technology refers to electronic devices that can be worn on the body, including smartwatches, fitness bands, smart eyewear, and health trackers. These devices use sensors and communication systems to gather information regarding the user's actions, health, and surroundings. AI integration involves incorporating artificial intelligence algorithms into these wearable devices. This enables devices to gather data and analyse them instantly, identify patterns, and offer actionable insights or feedback to the user. AI empowers wearables to handle and comprehend significant amounts of intricate data from sensors, including heart rate, movement, or environmental conditions. This results in insights that are more precise and significant [31].

Wearable technology consists of electronic accessories that can be donned on the body, including smartwatches, fitness bands, smart eyewear, and health trackers. These gadgets contain sensors and communication components to gather information on the user's actions, well-being, and surroundings. AI integration involves incorporating artificial intelligence algorithms into these wearable devices. This enables devices to gather data and analyse them instantly, identify patterns, and deliver actionable insights or feedback to the user. AI allows wearables to handle and make sense of extensive amounts of intricate data from sensors, including heart rate, movement, or environmental factors. This results in insights that are more precise and meaningful.

3.3.1 Managing Cardiovascular Disease with the Help of AI

With the capability of data integration, pattern recognition, and forecasting through the overall AI system, the availability of cardiac health monitors and wearable technologies has expanded due to AI advancements. Traditional CV evaluations employ a diverse, nonuniform, physician-driven method for collecting data that might overlook significant new symptoms or risks. However, a typical individual, such as a patient, can utilize wearable devices at any moment, such as a smart watch, an ECG patch, or garments embedded with biosensors to monitor ECG patterns, HRV, and various essential health metrics consistently [32]. The vast amounts of data can be efficiently utilized with advanced technologies such as artificial intelligence (AI), which assist in processing the information necessary for diagnosing cardiovascular diseases, predicting potential health issues, and providing straightforward health insights on the basis of a person's medical history. Artificial Intelligence Techniques for Heart Monitoring: Data gathered from wearables are analysed through various artificial intelligence methods, including a deep learning framework and traditional machine learning. These techniques examine ongoing, high-dimensional data flows from various sensors to detect anomalies and generate valuable insights. Machine learning algorithms, such as logistic regression, decision trees, and support vector machines, are utilized for tasks such as risk profiling and heartbeat pattern recognition. To classify new data from wearables and distinguish between normal and pathological rhythms, these models can be trained with labelled episodes of cardiac data.

3.3.2 Deep Learning Models

In tasks such as ECG waveform identification, both the CNN and RNN models achieve the greatest accuracy for intricate tasks. CNNs are effective in processing ECG readings, as they identify the smallest variations that may indicate an arrhythmia or other heart issues. As previously stated, the information gathered by wearables is in the format of time series data, making it ideal for analysis with RNNs. This additionally enhances monitoring and trends of heart health over time in people. Natural Language Processing (NLP): Various elements of NLP are utilized in cardiac monitoring for data analysis and interaction between the patient and the provider. To gain deeper insight into medical conditions, such as cases involving patient-reported data (symptoms), NLP might be employed alongside sensor data. AI Uses for Monitoring Cardiac Health: AI uses in wearable heart health monitoring encompass several significant research fields aimed at creating diagnostic and preventive healthcare interfaces. Detection of Arrhythmia: Specifically, the algorithms excel at identifying instances of arrhythmia, such as AFib. These models are capable of identifying patterns or even anomalies in the ECG and heart rhythm, which can be undetected without continuous monitoring. For example, deep learning models are beneficial in diagnosing AFib, which is quite challenging to identify in clinical settings because of its paroxysmal nature. Examination of Heart Rate Variability: Heart rate variability, known as HRV, serves as a valuable indicator of stress levels related to cardiac autonomic functions [33].

3.3.3 Cardiovascular Event Predictive Modelling

To calculate the likelihood of a cardiovascular event, algorithms analyse retrospective and concurrent data. For example, slight variations in a patient's blood pressure, HRV, or ECG over hours or days could signal the likelihood of developing new, intricate chronic conditions such as heart failure or myocardial infarction. In proactive healthcare, where early interventions can reduce the risk of serious cardiovascular events to a manageable level, such predictions are valuable. Health-related suggestions are generated by analysing users' health profiles and activity history, along with a statistical assessment of their baseline values. Their suggestions could involve altering one's lifestyle, modifying one's exercise routine, or advising a specific person to visit a doctor. As previously mentioned, heart health is quite distinct and is linked to actual behavior, environmental factors, and genetics. Consequently, this type of personalization is effective.

3.4 Predictive Analytics for Preventive Health

Predictive analytics is transforming numerous industries, and healthcare is included. It plays a crucial role in achieving success among teams, increasing patient care, and increasing operational efficiency for data-oriented organizations. This guide will delve into the significance of predictive analytics in healthcare, highlighting its advantages for various stakeholders and addressing potential challenges. Predictive analytics involves analysing current and past data via statistical modelling, data mining, AI, and machine learning methods to forecast upcoming events [34].

It pertains to various kinds of healthcare information, encompassing structured data, such as electronic health records (EHRs), and unstructured data, including clinical notes and imaging reports. It also utilizes big data to offer insights into patient care, resource management, and health outcomes. IT operators, care managers, and various healthcare professionals employ predictive analytics to reveal insights and facilitate operational efficiency [35]. Predictive modelling involves applying mathematical algorithms to data to forecast what is most likely to occur in the future.

3.4.1 Uses of Predictive Analytics in Healthcare

Predictive analytics in healthcare employs past patient data to anticipate future health patterns and results, allowing proactive measures and customized treatment. This technology aids in risk assessment, illness identification, treatment strategy development, and improvements in operational efficiency. The subsequent discussion focuses on the application of predictive analytics in the healthcare sector (Fig. 3.3).

I. **Readmission Prevention**

Approximately one in five adult patients face readmission to the hospital, so healthcare providers need to actively respond to patients' follow-up requirements through

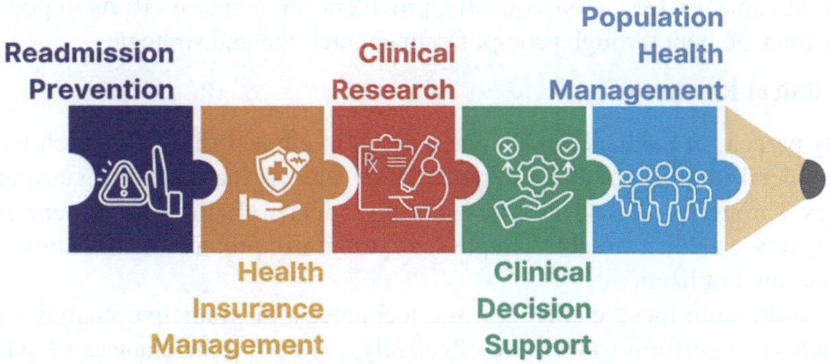

Fig. 3.3 Uses of predictive analytics in health care

tailored discharge plans. By utilizing predictive analytics in healthcare, providers can pinpoint patients vulnerable to readmission and distribute resources to avoid this situation. This prevention not only aids patients but also reduces healthcare system expenses related to readmissions, which currently total ~$52.4 billion annually in the U.S. It also helps 82% of hospitals affected by Medicare's Hospital Readmission Reduction Program (HRRP) [36].

II. **Clinical Decision Support**

Predictive analytics enables the anticipation of patient results, enhancement of treatment strategies, and elevation of overall care standards. Through the use of predictive analytics, healthcare professionals can detect early indicators of illnesses that may be hard to identify initially.

III. **Health Insurance Management**

Healthcare predictive analytics offers insights into the elements that affect care usage and expenses, which allows health insurance providers to gain a better understanding of the following: providers can actively address care requirements and reduce unnecessary expenses by utilizing predictive analytics. This enables them to precisely predict a patient's healthcare usage, including hospital admissions or surgeries, by utilizing historical data and clinical factors [37].

Forecasting a patient's anticipated healthcare expenses can be achieved by analysing their previous usage and treatment requirements. Fraud prevention: Payers can examine claims data and spot patterns of fraudulent behavior to detect fraud early and avert insurance misuse. This would greatly affect the financial aspect of patient care, as losses caused by healthcare fraud represent 3–10% of healthcare expenditures, totaling $300 billion. Determining the precise cost of insurance enables patients to obtain the necessary care without financial obstacles.

IV. **Management of Population Health**

Providers can analyse patient data beyond individual treatment to pinpoint groups with common medical traits and avert health risks at the population level. Providers might identify disease outbreaks sooner by using predictive analytics to spot groups at risk of exposure [38]. Consequently, providers can increase efforts in population health management through prompt treatment creation and strategies.

V. **Clinical Research**

Ninety percent of medications do not succeed in clinical trials. Research indicates that artificial intelligence has the potential to address the problems causing these failures. Through predictive analytics, AI can recognize eligible patients, enroll likely participants, establish suitable sample sizes, minimize protocol inconsistencies, and increase fund utilization.

By addressing these challenges, the technology of predictive analytics allows researchers to perform trials more effectively, increasing the chances of achieving innovative outcomes.

3.4.2 Challenges and Ethical Considerations

Although it offers numerous advantages, predictive analytics in healthcare also presents risks and challenges. Key concerns include the following: (i) Data security and privacy: Due to the sensitive characteristics of health information, healthcare data are a significant target for cybersecurity threats. To maintain patient confidence, adhere to the HIPAA guidelines, and safeguard individual privacy, healthcare providers need to implement rigorous data protection protocols. (ii) Precision and dependability: Predictive models face ongoing examination to prevent possible misdiagnoses, biases, or treatment mistakes. This results in diminished confidence in the model. (iii) Provider adoption: Providers need to find a balance between improving patient care via data gathering and the time and effort needed to utilize analytics. Training on emerging technologies and adopting new workflows may lead to considerable adjustments for providers. This does not automatically mean the end for predictive analytics in healthcare. By implementing data access controls and securing consent prior to data sharing, providers can leverage predictive analytics while ensuring that patient privacy is safeguarded. Users must still approach any technology with a clear understanding of when it is suitable, how to prevent errors, and how to critically analyse results. Ongoing attention is also necessary for ethical considerations such as addressing healthcare inequalities and reducing bias in data and algorithms. For example, implicit biases may infiltrate predictive analysis algorithms, resulting in unjust outcomes.

Predictive analytics influences all areas of healthcare and creates thrilling opportunities, ranging from timely interventions to more personalized, considerate, and interactive patient experiences. As a potent instrument for improving healthcare services, it provides insights that are crucial for fostering efficiency, precision, and better patient results. The prospective path of predictive analytics in healthcare is equally encouraging and demanding. Innovative technologies such as AI greatly enhance predictive analytics functions, and healthcare organizations need to adopt this technology to remain leaders in healthcare advancement. Achieving success in predictive analytics necessitates a solid foundation, and an advanced architecture (such as a lakehouse) is the initial step in utilizing this technology effectively [39]. The incorporation of these technologies into healthcare processes is essential for maximizing their benefits and enhancing patient outcomes and operational effectiveness. By combining thoughtful considerations with innovative ideas, advocates of technology and healthcare trailblazers can work together to create a more predictive and proactive healthcare system.

3.5 Conclusions

The incorporation of AI technologies into wellness systems signifies a crucial change in the ways in which people, organizations, and communities view health and well-being. Through predictive analytics that foresee health risks to tailor suggestions that enhance lifestyle decisions, AI enables individuals to actively manage their physical, mental, and emotional well-being. The utilization of information from various sources—such as wearable technology, health records, genetic data, and environmental influences—AI facilitates a comprehensive and integrated method for health management. Nonetheless, this change has gone further than just the technology alone. The true strength of AI is its capacity to promote a more human-focused wellness culture—prioritizing prevention instead of treatment, personalization rather than standardization, and accessibility in place of exclusivity. AI-powered wellness solutions can increase productivity, lower healthcare expenses, and foster a more engaged and resilient workforce for organizations. However, along with this promise comes accountability. The design of ethical AI, data protection, openness, and fair access should remain at the forefront to guarantee that these improvements benefit everyone without worsening current disparities. As AI advances, its function in wellness progressively transitions from a supportive resource to a fundamental ally in promoting healthier lives and communities. Ultimately, the change led by AI is not just about technology redefining wellness—it is about redefining wellness itself. By merging human compassion with artificial intelligence, we can create a future where well-being is not an exclusive benefit but rather a reality accessible to everyone.

References

1. Danda, R. R., Yasmeen, Z., & Maguluri, K. K. (2020). *AI-driven healthcare transformation: Machine learning, deep learning, and neural networks in insurance and wellness programs.* JEC PUBLICATION.
2. Bhatt, N., Bisht, B., Mittal, A., & Garg, N. (2024). Tech-driven wellness: Navigating the confluence of technology and digital health. In *Proceedings of the 2024 2nd international conference on artificial intelligence and machine learning applications theme: Healthcare and internet of things (AIMLA)* (pp. 1–6). IEEE.
3. Ganesan, A. (2025). Transforming wellness: Ethical, cognitive, and privacy challenges in AI-driven healthcare. In *Harnessing AI and machine learning for precision wellness* (pp. 481–500). IGI Global Scientific Publishing.
4. Dhinakaran, D., Raja, S. E., Jasmine, J. J., Kumar, P. V., & Ramani, R. (2025). The future of well-being: AI-powered health management with privacy at its core. *Wellness Management Powered by AI Technologies, 16*, 363–402.
5. Joshi, P. T., Tsabary, E., & Gamage, I. P. (2025). AI and language: Transforming communication in health and wellness. In *Driving global health and sustainable development goals with smart technology* (pp. 181–212). IGI Global Scientific Publishing.
6. Mittal, S., Wijethilak, H. P., & Chandel, A. (2025). Effectiveness of AI-powered wellness in promoting employee wellbeing and happiness. In *Insights on wellbeing and happiness in the workplace* (pp. 87–108). IGI Global Scientific Publishing.

7. Gkintoni, E., Telonis, G., Halkiopoulos, C., & Boutsinas, B. (2024). Next-gen wellness: AI, MBCT, and data-driven marketing transforming health tourism. In *International conference of the international association of cultural and digital tourism* (pp. 819–855). Springer.
8. Gkintoni, E., Telonis, G., Tsimakis, A., Halkiopoulos, C., & Boutsinas, B. (2024). Integrating AI-powered personalization via neurobiologically informed DBT in smart wellness resorts. In *International conference of the international association of cultural and digital tourism* (pp. 923–953). Springer.
9. Rao, T. V. N., Patel, G. S., Bhavana, S., & Swetha, C. (2025). AI-driven decision support system for workplace wellbeing. In *Insights on wellbeing and happiness in the workplace* (pp. 1–18). IGI Global Scientific Publishing.
10. Tiwari, M., & Waoo, A. A. (2024). For intelligent, personalized well-being. *Revolutionizing Healthcare: AI Integration with IoT for Enhanced Patient Outcomes, 7*, 109.
11. Agarwal, P., Swami, S., & Malhotra, S. K. (2024). Artificial intelligence adoption in the post COVID-19 new-normal and role of smart technologies in transforming business: A review. *Journal of Science and Technology Policy Management, 15*(3), 506–529.
12. Majumder, S., Chowdhury, S., Dey, N., & Santosh, K. C. (2022). Balance your work–life: Personal interactive web-interface.
13. Rahman, M. A., Victoros, E., Ernest, J., Davis, R., Shanjana, Y., & Islam, M. R. (2024). Impact of artificial intelligence (AI) technology in healthcare sector: A critical evaluation of both sides of the coin. *Clinical Pathology, 17*, 2632010.
14. Puaschunder, J. M. (2019). Artificial intelligence in the healthcare sector. *Scientia Moralitas-International Journal of Multidisciplinary Research, 4*(2), 1–14.
15. Saxena, A. K., Ness, S., & Khinvasara, T. (2024). The influence of AI: The revolutionary effects of artificial intelligence in healthcare sector. *Journal of Engineering Research and Reports, 26*(3), 49–62.
16. Pagallo, U., O'Sullivan, S., Nevejans, N., Holzinger, A., Friebe, M., Jeanquartier, F., et al. (2024). The underuse of AI in the health sector: Opportunity costs, success stories, risks and recommendations. *Health and Technology, 14*(1), 1–14.
17. Palaniappan, K., Lin, E. Y. T., & Vogel, S. (2024). Global regulatory frameworks for the use of artificial intelligence (AI) in the healthcare services sector. *Healthcare, 12*(5), 562.
18. Iqbal, S., Tariq, M., Ayesha, H., & Ayesha, N. (2021). AI technologies in health-care applications. In *Artificial intelligence and internet of things* (pp. 3–44). CRC Press.
19. Das, S., Chakraborty, S., Sajjan, G., Majumder, S., Dey, N., & Tavares, J. M. R. (2022). Explainable AI for predictive analytics on employee attrition. In *International conference on soft computing and its engineering applications* (pp. 147–157). Springer.
20. Babar, Z., Paul, R., Barua, T., & Rahman, M. A. (2025). AI applications in emerging tech sectors: A review of AI use cases across healthcare, retail, and cybersecurity. *ASRC Procedia: Global Perspectives in Science and Scholarship, 1*(01), 16–33.
21. Mennella, C., Maniscalco, U., De Pietro, G., & Esposito, M. (2024). Ethical and regulatory challenges of AI technologies in healthcare: A narrative review. *Heliyon, 10*(4), 1334.
22. Alhashmi, S. F., Alshurideh, M., Al Kurdi, B., & Salloum, S. A. (2020). A systematic review of the factors affecting the artificial intelligence implementation in the health care sector. In *The international conference on artificial intelligence and computer vision* (pp. 37–49). Springer.
23. Kumar, A., & Joshi, S. (2022). Applications of AI in healthcare sector for enhancement of medical decision making and quality of service. In *Proceedings of the 2022 international conference on decision aid sciences and applications (DASA)* (pp. 37–41). IEEE.
24. Alhashmi, S. F., Salloum, S. A., & Mhamdi, C. (2019). Implementing artificial intelligence in the United Arab Emirates healthcare sector: An extended technology acceptance model. *International Journal of Information Technology and Language Studies, 3*(3), 4985.
25. Shaheen, M. Y. (2021). Applications of artificial intelligence (AI) in healthcare: A review. *ScienceOpen Preprints, 32*, 4521.
26. Graham, S., Depp, C., Lee, E. E., Nebeker, C., Tu, X., Kim, H. C., & Jeste, D. V. (2019). Artificial intelligence for mental health and mental illnesses: An overview. *Current Psychiatry Reports, 21*(11), 116.

27. Majumder, S., Ghosh, S., & Polkowski, Z. (2023). A brief introduction to human resource management. In *Artificial Intelligence Techniques in Human Resource Management* (pp. 1–25). Apple Academic Press.
28. D'Alfonso, S. (2020). AI in mental health. *Current Opinion in Psychology, 36*, 112–117.
29. Lovejoy, C. A. (2019). Technology and mental health: The role of artificial intelligence. *European Psychiatry, 55*, 1–3.
30. Cross, S., Bell, I., Nicholas, J., Valentine, L., Mangelsdorf, S., Baker, S., et al. (2024). Use of AI in mental health care: Community and mental health professionals survey. *JMIR Mental Health, 11*(1), e60589.
31. Luxton, D. D. (Ed.). (2015). Artificial intelligence in behavioral and mental health care.
32. Huang, S., Lai, X., Ke, L., Li, Y., Wang, H., Zhao, X., et al. (2024). AI technology panic—is AI dependence bad for mental health? A cross-lagged panel model and the mediating roles of motivations for AI use among adolescents. *Psychology Research and Behavior Management, 52*, 1087–1102.
33. Luxton, D. D., Anderson, S. L., & Anderson, M. (2016). Ethical issues and artificial intelligence technologies in behavioral and mental health care. In *Artificial intelligence in behavioral and mental health care* (pp. 255–276). Academic Press.
34. Mittal, A., Dumka, L., & Mohan, L. (2023). A comprehensive review on the use of artificial intelligence in mental health care. In *Proceedings of the 2023 14th international conference on computing communication and networking technologies (ICCCNT)* (pp. 1–5). IEEE.
35. Ghosh, S., Majumder, S., & Peng, S. L. (2023). An empirical study on adoption of artificial intelligence in human resource management. In *Artificial intelligence techniques in human* resource management (pp. 29–85). Apple Academic Press.
36. Khosravy, M., Gupta, N., Dey, N., & Witkowski, O. (2025). Are LLMs ready to shoulder biomedical trust? In *International conference on innovations in biotechnology and health sciences* (pp. 125–142). Springer.
37. Huyen, D. T. T., Binh, N. T., Tuan, T. M., Trung, T. Q., Nhu, N. G., Dey, N., & Son, L. H. (2017). Analyzing trends in hospital-cost payments of patients using ARIMA and GIS: Case study at the Hanoi Medical University Hospital, Vietnam. *Journal of Medical Imaging and Health Informatics, 7*(2), 421–429.
38. Maciejewski, R., Hafen, R., Rudolph, S., Larew, S. G., Mitchell, M. A., Cleveland, W. S., & Ebert, D. S. (2010). Forecasting hotspots—a predictive analytics approach. *IEEE Transactions on Visualization and Computer Graphics, 17*(4), 440–453.
39. Abd-Alrazaq, A., Alhuwail, D., Schneider, J., Toro, C. T., Ahmed, A., Alzubaidi, M., et al. (2022). The performance of artificial intelligence-driven technologies in diagnosing mental disorders: An umbrella review. *NPJ Digital Medicine, 5*(1), 87.

Chapter 4
AI in Workplace Wellness Programs

AI in workplace wellness initiatives involves the use of artificial intelligence tools to improve and customize employee health and wellness programs. These technologies evaluate data, forecast possible health hazards, and offer personalized suggestions to enhance employee well-being. AI can customize wellness initiatives, provide real-time assistance, monitor health information, and anticipate possible burnout. It can evaluate employee information to pinpoint health risks and predict future health issues [1]. For example, AI can detect indicators of stress or burnout by examining work habits, productivity rates, and biometric information. This enables employers to act proactively and deliver support before issues worsen. AI algorithms evaluate employee information (e.g., health records, lifestyle choices, social media usage) to develop personalized wellness strategies. This may involve tailored workout programs, dietary strategies, and methods for managing stress. Platforms powered by AI can recommend activities and resources tailored to personal preferences and requirements.

4.1 Encouraging Employee Engagement and Productivity

To enhance employee engagement and productivity, establishing a supportive work atmosphere, encouraging transparent dialogue, offering development opportunities, acknowledging accomplishments, and advocating for work–life balance are emphasized. Managers should model behavior, provide assistance, and proactively seek and respond to employee input. Distinct and regular communication pathways, incorporating both formal and informal conversations, should be created to guarantee that employees sense that they are acknowledged and appreciated. Employee engagement reflects the level of an employee's connection to their company, directly affecting their motivation to excel in their job. On the other hand, it might also be referred to as a strategy that a company develops to foster an atmosphere where employees

S. Majumder and N. Dey, *AI-Driven Wellbeing to Enhance Lives and Work Environments*, Studies in Computational Intelligence 1257,
https://doi.org/10.1007/978-3-032-15735-5_4

are motivated to realize more of their capabilities. Consequently, there is a positive correlation between employee engagement and productivity [2].

Employee engagement has emerged as a catchword in management terminology in recent years [3]. Each organization recognizes the value of a dedicated, proactive worker rather than one who simply fulfils their duties and develops employee engagement strategies accordingly.

Unlike sales revenue, which is difficult to measure, however, its impact permeates every area of the organization, influencing employee satisfaction, performance, and profits.

Figure 4.1 depicts a circular connection among three essential business results. Employees who are driven, dedicated, and emotionally invested in their tasks typically achieve greater performance. This increased involvement leads to increased productivity. Increased efficiency and improved performance allow the organization to produce more with the same or fewer resources, resulting in higher profitability. Enhanced profits offer the means and motivation to reinvest in employees, providing superior benefits, training, and work environments, which subsequently enhances employee engagement. The arrows in the triangle indicate that this forms a constant, reinforcing loop—each element positively impacts the next, fostering a sustainable cycle of growth and achievement [4]. AI can greatly enhance employee engagement and productivity by tailoring experiences, automating tasks, and delivering important insights (Fig. 4.2).

The workplace influences employee experience. It encompasses culture, the physical environment, and accessible resources. An uplifting atmosphere promotes enthusiasm and contentment. Effective communication, chances for collaboration, and inviting environments are crucial. A positive environment results in increased productivity, enhanced collaboration, and a dedication to organizational objectives. Organizational culture consists of a collective set of values and practices within a company. It affects employee motivation and performance. A robust culture fosters togetherness and confidence. When workers resonate with the company's mission, they perform better and are more inclined to stay with the organization. Leaders influence the employee experience by providing transparent communication and genuine assistance. Consistent feedback fosters trust and encourages motivation. Effective

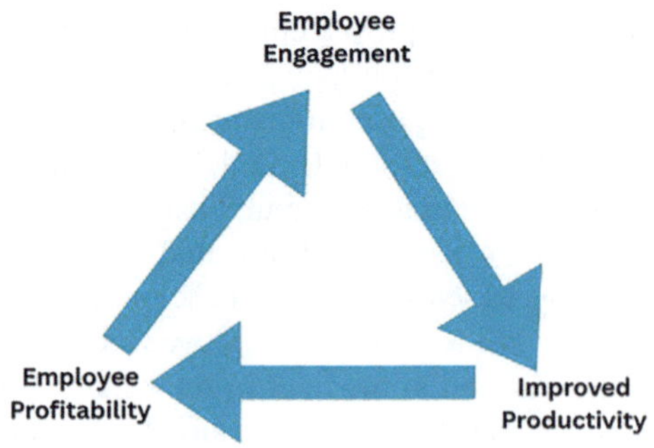

Fig. 4.1 Conceptual framework employee engagement and productivity

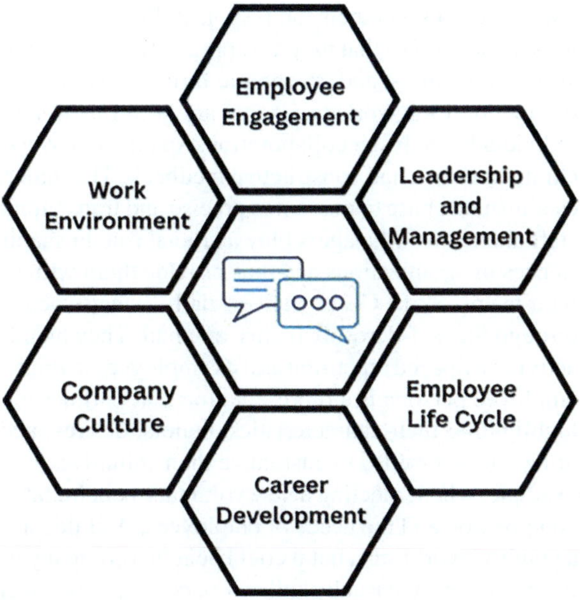

Fig. 4.2 Key components of the employee experience

managers acknowledge employees' contributions and offer direction. Setting achievable goals and expectations fosters job satisfaction and development. Effective leadership motivates dedication to organizational achievement, fostering beneficial results for both individuals and groups. The engagement of employees is essential for an efficient work environment. This guarantees that employees are engaged and linked to their positions. Employees who are engaged demonstrate greater satisfaction and eagerness to contribute. They cultivate a collaborative environment [5]. Fostering robust engagement aids in talent retention and contributes to organizational development. Workers aim to improve abilities and advance in their careers. Training initiatives, seminars, and guidance promote development. Offering educational choices demonstrates dedication to employees. Workers who envision their future in their positions remain committed and engaged, enhancing the success of the organization. The employee lifecycle includes recruiting, onboarding new hires, development, retention, and departure. Every phase of the employee life cycle influences an employee's path and experience. An effectively managed life cycle enhances employee contentment and leads to organizational achievement [6]. Smooth transitions and effective communication are emphasized to sustain motivation and dedication during the entire process.

Enhancing the employee experience is an ongoing endeavor. Regularly listening to and comprehending employee sentiment is essential. Regularly conducting employee experience surveys enables us to determine how employee needs evolve (Fig. 4.3). Well-being programs aimed at physical, mental, and emotional wellness can alleviate stress and enhance health results and job satisfaction. Providing gym memberships, mental health days, meditation sessions, or virtual therapy shows concern for employees beyond their job responsibilities. Offering resources for employee wellness demonstrates that they are appreciated beyond the workplace [7]. Contemporary spaces, amenities, and ergonomic furnishings in office settings foster creativity and ensure comfort. Workspaces that are both functional and visually attractive inspire individuals and foster collaboration. An effective work atmosphere depends on strong communication and constructive feedback. The majority of successful organizations establish this base to guarantee precise and transparent information exchange among staff and teams. Managers play a crucial role in shaping employees' everyday experiences in organizations and thus provide them with the resources necessary to assist their teams. Skills Coach delivers daily, concise activities and educational resources through Slack, Microsoft Teams, or email. They blend effortlessly into their everyday activities. Spaced repetition assists employees in improving their leadership skills. An employee persona is a representation that outlines a particular group of employees, highlighting their characteristics, beliefs, desires, and requirements. HR teams can utilize these profiles to customize their initiatives for various employee groups. For example, when selecting a new voluntary benefit, take into account an organization's composition and the needs of employees. A childcare allowance may not be best for a youthful workforce, but it could benefit a company with numerous caregivers. Even if one currently casually utilizes personas, it is advantageous to create and record 3–5 personas for a job. In this manner, new HR team members can conveniently refer to them when new initiatives are initiated, guaranteeing that various employee viewpoints have been taken into account. Offer training sessions, workshops, or mentoring programs for employees to enhance skills and progress in their careers [8]. Companies must allocate resources to employee growth to achieve a competitive edge. Workers with chances to learn are more inclined to remain loyal and put in extra effort. Professional growth fosters creativity, enabling employees to discover new approaches to overcome challenges. Investing in development enhances workplace culture, fostering a positive atmosphere where employees feel appreciated and esteemed. Technology and tools streamline activities and increase productivity. Intuitive software, automation applications, and online platforms simplify tasks. Starting by pinpointing aspects that need enhancement when developing an employee experience framework. Lists to employee input and aligns strategies with organizational objectives. Set specific goals and assess the results. The use of instruments and materials that foster a supportive environment motivates employees to excel and engage efficiently. The transition to remote and hybrid work necessitates companies to reconsider their adaptability. A 2021 E&Y study revealed that 22% of workers favour full-time office work, whereas 67% believe that productivity can be assessed irrespective of location. Maintaining uniformity between home and office personnel is essential. Remote employees frequently sense a lack of connection or face-to-face interactions and advancement opportunities and face a "belonging tax." Organizations need to convey messages effectively, organize virtual social gatherings, educate hybrid managers, and refrain from prioritizing in-office staff to close this gap. Performance management systems can be utilized to monitor and evaluate employee accomplishments. Establish clear expectations and consistently offer constructive feedback. We acknowledge the achievements and pinpoint areas for improvement. Managing

Fig. 4.3 Prioratize employee experience

performance can enhance motivation and assist employees in grasping their responsibilities. Promote establishing goals that correspond with organizational aims and enhance responsibility. Managers need to receive training in inclusive leadership practices. Research indicates that employees perform and engage more effectively when they appreciate their diversity, are heard, are involved in decision-making, are supported, and are able to be their authentic selves. Various viewpoints help employees feel valued and enhance morale. This method also fosters a positive and trusting workplace environment.

The contemporary work environment is characterized by speed and heightened digital integration, making employee involvement and knowledge exchange essential for an organization's success [9, 10]. Businesses continuously seek innovative and imaginative methods to sustain employee involvement and promote knowledge exchange among team members. Artificial intelligence (AI) has emerged as a powerful resource that can be crucial for achieving these goals. This study revealed that artificial intelligence technologies offer managers essential insights and data-driven information to assist them in making decisions about their team members. Artificial intelligence (AI) can analyse employee feedback from multiple sources, such as surveys, performance evaluations, and communication channels, to identify patterns, problems, and areas for improvement. Methods for natural language processing (NLP) can be utilized to derive themes and emotions from feedback. Managers can utilize these data to address specific issues and make informed, data-driven choices. This discovery aligns with the results of the present study. Employee engagement and AI. AI influences knowledge sharing and employee engagement in organizations in both beneficial and detrimental ways [11]. Nonetheless, AI-driven tools can provide personalized feedback, coaching, and training initiatives that enhance employee skills and knowledge, resulting in increased engagement levels. Organizations must employ AI in a responsible, transparent manner while honoring the privacy and autonomy of their employees to mitigate these risks. Organizations can deploy AI-driven virtual assistants and decision support systems to grant employees more independence and authority in their roles, thereby enhancing engagement through artificial intelligence (AI). By utilizing coaching programs and

AI-driven learning platforms, they can also offer increased opportunities for development and education. Organizations can utilize AI-driven communication tools that summarize long documents or emails into concise formats and automatically translate messages into various languages according to the recipient's preferences, enhancing staff communication. Organizations can implement AI-driven systems that rank tasks on the basis of priority and urgency and suggest potential collaborators on the basis of their expertise and knowledge to enhance effective teamwork among employees [12]. Overall, artificial intelligence (AI) holds significant promise for assisting organizations in thriving within the rapidly evolving digital environment by promoting knowledge sharing and enhancing employee involvement. Artificial intelligence tools offer innovative methods for personalizing engagement strategies and enhancing the efficient transfer of information within the workplace. Onboarding enhanced by AI, chatbots for assistance, and tailored feedback all increase employee engagement by providing customized experiences and continuous support. The deployment of artificial intelligence necessitates the creation of defined ethics and governance protocols within organizations, involving employees in conversations about the use of artificial intelligence and regularly auditing and assessing artificial intelligence systems for accountability, fairness, and transparency. The application of artificial intelligence in knowledge sharing and employee engagement can become more ethical and responsible by adhering to ethical artificial intelligence principles [13]. Artificial intelligence can retrieve and condense information from unstructured data sources such as emails, social media, and research papers. This function simplifies the collection and sharing of standard protocols and knowledge. Moreover, it can help convert implicit information into explicit knowledge that is simpler to disseminate throughout the organization. The automation driven by artificial intelligence could simplify knowledge exchange processes. Routine tasks in knowledge management can be automated through robotic process automation (RPA), allowing employees to focus on more valuable work. This boosts efficiency and ensures that knowledge-sharing initiatives are implemented regularly. AI-driven recommender systems play a crucial role in disseminating information. They improve the visibility of knowledge resources by examining user behavior and preferences to suggest pertinent information. AI personalization algorithms enhance knowledge-sharing platforms by tailoring content to individual users' needs and preferences, thereby increasing user engagement and knowledge retention. Artificial intelligence can enhance knowledge sharing by providing smart tools and platforms that increase the effectiveness, accuracy, and efficiency of information transfer processes [14]. Language, location, and time constraints are only some of the challenges that artificial intelligence can help eliminate to facilitate effective information sharing. Using artificial intelligence in knowledge exchange has both benefits and drawbacks (Fig. 4.4).

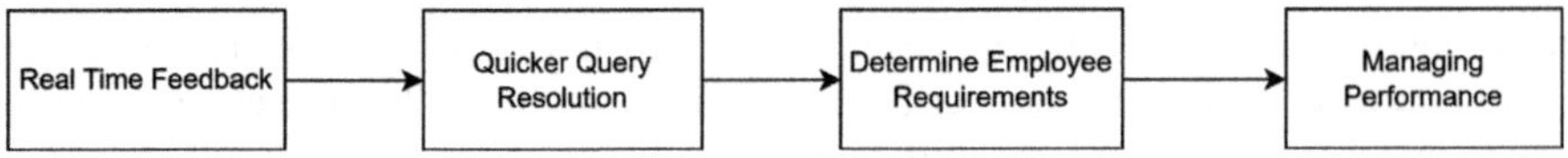

Fig. 4.4 Improving employee engagement with AI

4.2 Stress Management and Mental Health Initiatives

Stress is a worldwide concern that has significant impacts on mental and physical well-being. As awareness of its effects increases, the demand for efficient tools to identify and handle stress is increasing. The human-stress-detection system meets this requirement by combining cutting-edge machine learning methods with real-time AI-powered assistance. This system uses a random forest classifier algorithm to evaluate physiological data, including snoring frequency, breathing rate, and heart rate, to precisely forecast stress levels. By integrating an AI assistant, the system delivers customized and interactive support via chat, boosting user involvement and offering personalized stress management suggestions. This document investigates the creation and capabilities of the human-stress-detection system, outlining its structure, data processing techniques, and the efficiency of its AI-powered functionalities. Through the integration of machine learning with immediate interaction, the system seeks to deliver an all-encompassing approach to stress management, offering significant perspectives to both the area of mental health and the wider research community [15, 16].

4.2.1 Evolution of Mental Health and Stress Management

The evolution of mental health and stress management has been considerable over the years, mirroring shifts in societal perceptions, scientific studies, and therapeutic methods: (i) Initial Comprehension: In the past, mental health problems were frequently misinterpreted and associated with stigma. Ancient societies believed that supernatural powers or moral shortcomings caused mental illnesses. Treatments are basic and frequently include severe approaches such as exorcisms or isolation. (ii) Nineteenth century progress: the nineteenth century signaled the start of a more scientific methodology toward mental health. The creation of asylums and the emergence of psychiatry offered more compassionate and organized treatment. Sigmund Freud's advancement of psychoanalysis and the emergence of psychological theories established the foundation for contemporary mental health practices. (iii) Twentieth century advances: the twentieth century saw significant advancements in the understanding of mental health and stress. The development of psychotropic medications has revolutionized treatment, whereas behavioural therapies and cognitive-behavioural therapies (CBTs) have offered new methods for tackling mental health challenges. Research into stress as a psychological and physiological phenomenon has gained traction, leading to a deeper understanding of its impact on overall health. Modern Methods: In recent years, there has been an increasing focus on holistic and integrative strategies for mental health and stress management. Technological advancements have brought forth new instruments for identifying and managing stress, such as wearable gadgets and AI-based systems. (iv) Present trends: At this time, mental well-being and stress reduction remain fields of ongoing investigation and development. The incorporation of machine learning and artificial intelligence into mental health services signifies a major advancement, presenting fresh opportunities for immediate tracking and tailored assistance. Current research aims to deepen our understanding of mental health and create effective methods for coping with stress in a fast-evolving environment [17, 18].

4.2.2 AI and ML Contributions to Stress Management

Artificial intelligence (AI) and machine learning (ML) are advancing rapidly in stress management by providing advanced tools for detection, prediction, and intervention. These technologies utilize intricate algorithms to evaluate extensive volumes of physiological and behavioral data, thus improving the precision and performance of stress management techniques. AI systems utilize sophisticated methods such as random forest classifiers and neural networks to analyse data, including heart rate variability, snoring frequency, and sleep patterns, enabling tailored stress forecasts. This feature enables the development of personalized stress profiles, providing insights customized to the distinct physiological and psychological traits of each user. Real-time observation is among the most significant benefits of AI and ML in managing stress. Wearable devices and intelligent sensors featuring AI technology consistently monitor physiological signals such as heart rate and skin conductivity. This ongoing observation offers users prompt insights into their stress levels, allowing them to identify and manage stressors in real time. The ability to deliver real-time insights allows users to adopt coping strategies quickly, which is vital for successful stress management. This prompt feedback facilitates a proactive strategy for handling stress, lessening its effects before it intensifies.

AI and ML are essential for providing tailored suggestions for managing stress [19]. Machine learning models assess personal stress patterns and past data to create tailored intervention strategies. These suggestions could involve particular relaxation methods, cognitive-behavioral practices, or lifestyle changes tailored to the user's stress profile. This tailored method improves the significance and efficiency of the suggestions, facilitating users in embracing and maintaining stress management techniques. By customizing guidance to meet the unique requirements of individuals, AI-based systems offer more valuable assistance in handling stress. The accessibility and interaction offered by AI-driven platforms, including virtual assistants and chatbots, represent another significant progress in stress management. These interactive tools provide immediate assistance and direction, facilitating user engagement with stress management resources [20]. AI assistants can deliver immediate answers to user questions, suggest coping mechanisms, and lead users through stress-relief activities. This degree of accessibility enhances user involvement and commitment to stress management practices, as individuals can obtain assistance whenever necessary, free from the limitations of conventional therapy or support meetings. Additionally, the function of AI and ML in examining stress patterns among populations provides essential insights for researchers and healthcare professionals. Through the analysis of extensive datasets, AI can detect general patterns and trends in stress levels, aiding in a deeper comprehension of stress-related problems at a societal scale. This knowledge supports the creation of focused interventions and health initiatives that address stress-related issues more efficiently.

4.2.3 Alignment of Mental Health with Artificial Intelligence

Artificial intelligence has become a groundbreaking influence across multiple domains, and its use in mental health care is no different. Thus, this review examines the incorporation of AI into mental health care, clarifying existing trends, ethical concerns, and future pathways in this evolving domain. This review included recent research, instances of AI usage, and ethical factors influencing the domain. Furthermore, regulatory structures and patterns in research and development have been examined [20]. Four databases (PubMed, IEEE Xplore, PsycINFO, and Google Scholar) were explored thoroughly. The criteria for inclusion were articles published in peer-reviewed journals, conference proceedings, or respected online databases; studies that concentrate specifically on the utilization of AI in mental healthcare; and review articles that provide a thorough overview, analysis, or synthesis of previously published literature in English. Ongoing trends demonstrate AI's groundbreaking capabilities, featuring uses such as early identification of mental health issues, customized treatment strategies, and AI-powered virtual therapists [21]. Nonetheless, these developments present ethical dilemmas related to privacy, bias reduction, and maintaining the human aspect of therapy. Future pathways highlight the necessity for well-defined regulatory structures, open validation of AI models, and ongoing research and development initiatives. The incorporation of AI into mental health treatment and mental healthcare signifies an encouraging new direction in the medical field. Although AI can transform mental healthcare, ensuring its responsible and ethical application is crucial. By addressing these obstacles and carefully guiding future paths, we can effectively harness AI's ability to improve the accessibility, effectiveness, and ethics of mental health services, benefiting both individuals and communities [22].

AI can monitor internet activities and social media interactions to identify indicators of cyberbullying, anxiety, or depression in young individuals. AI-driven early intervention can foster healthy emotional coping strategies and prevent the worsening of mental health problems. One straightforward but impactful technique for stress management involves the use of AI-based mindfulness apps. These tools examine user habits and preferences, offering customized meditation sessions, stress-reduction activities, and cognitive behavioral therapy methods.

AI is increasingly utilized in stress management by offering personalized assistance, detecting stress signals early, and automating tasks that induce stress. AI is capable of examining multiple data indicators, such as facial expressions, tone of voice, and biometric information, to identify stress, subsequently offering tailored interventions such as mindfulness practices or recommending lifestyle changes [23].

Wearable devices with AI capabilities can track heart rate, sleep habits, and activity levels to identify initial indicators of stress. AI can assess facial cues, vocal tone, and breathing rhythms to recognize stressors and trends. AI is capable of assessing social media engagement, usage duration, and digital behavior to detect possible signs of stress. AI can monitor posture and different biometric data to identify potential stressors in the workplace and can evaluate user behavior to recommend adjustments to daily routines, sleep habits, or work practices to increase stress resilience. AI can offer prompt reminders for taking breaks, engaging in relaxation methods, or obtaining professional assistance when necessary. AI aids in prioritizing tasks, distributing workloads, and automating monotonous activities, minimizing feelings of being overburdened. AI can evaluate different data points to detect initial indicators of anxiety, depression, or burnout, facilitating prompt interventions [24]. AI can link users with mental health experts, support networks, or pertinent resources tailored to their specific requirements. AI can offer advice on sleep hygiene, physical activity, and various other healthy lifestyle options that help reduce stress. AI is changing our methods of managing stress by providing tailored, proactive, and preventive solutions. By utilizing AI's potential, people and companies can enhance their comprehension and control and ultimately diminish the adverse effects of stress on their well-being.

4.2.4 AI Can Help with Stress Management at Work

AI-based tools can be utilized in various industries to assist employees in handling stress more efficiently. There are several methods through which AI is utilized for managing stress in the workplace:

I. Tailored meditation and relaxation: Applications such as HeadSpace and Calm employ AI algorithms to adapt meditation and relaxation methods according to a user's stress behaviors and interaction history [25].
II. AI for productivity and organization: Tools powered by AI, such as Notion or Microsoft Copilot, assist employees in maintaining organizations, managing tasks effectively, and prioritizing work to alleviate stress. AI-powered scheduling tools such as Motion or Clockwise enhance calendars to avoid overwhelm and maintain work–life harmony [26].
III. AI stress monitors in the workplace: Certain organizations are implementing AI-based analytical tools that evaluate employee workloads, identify burnout potential, and suggest proactive measures such as recommending rest periods or modifying workloads.
IV. AI-integrated wearables: Smartwatches and biometric devices such as Fitbits utilize AI to track stress levels by analysing heart rate variability and recommend relaxation methods when stress increases. Although AI can be an effective aid for managing stress, it should not serve as a substitute for professional mental health services. Therapists, medical professionals, and support groups are vital

for holistic wellness. AI can act as an extra support system, providing guidance and resources when necessary, but professional assistance should be pursued for significant mental health issues. For additional insights into ethical AI applications in mental health, watch Stevie Chancellor's TED Talk on Creating AI to Enhance Mental Health [27].

AI transforms how we handle stress, providing tailored, accessible, and data-informed approaches. Technology simplifies the process of managing stress and improving well-being through AI-based therapy, mindfulness apps, workplace stress tracking, and additional tools [28, 29]. Productivity tools driven by AI can assist in maintaining organizations, alleviating workload stress, and enhancing efficiency—factors that contribute to a better work–life balance. Nonetheless, AI must not take the place of therapists, healthcare professionals, or support networks. AI should function as a supplementary support system, enhancing traditional stress management and mental health tools. With the ongoing advancement of AI, its significance in stress management will increase, assisting individuals from various professions and lifestyles in flourishing collectively.

4.3 AI-Driven Fitness and Nutrition Programs

AI-powered fitness and nutrition programs leverage artificial intelligence to customize exercise and meal plans, monitor progress, and provide individualized support. These programs utilize data analysis and machine learning to cater to personal needs and preferences, enhancing accessibility and effectiveness in fitness and nutrition. The fitness sector is changing as AI in fitness revolutionizes how companies interact with users and enhance their operations. Generic services do not satisfy current user expectations, and companies must adapt to remain competitive [30]. With the increasing link between fitness and artificial intelligence, brands are discovering more intelligent methods to customize experiences and enhance outcomes.

AI is revolutionizing all aspects, from customer interaction to operational productivity, offering fitness brands a chance to dominate the market. AI allows companies to deliver highly personalized exercise regimens on the basis of user objectives, tastes, and real-time metrics. For example, Freeletics employs AI algorithms to modify plans dynamically, enhancing user retention and loyalty as they progress. By utilizing AI fitness app solutions, the application will offer every user a personalized fitness journey, ensuring sustained engagement over time. AI in fitness apps offers immediate insights into user behavior, health metrics, and trends. Apps such as Fitbit evaluate data to provide practical advice, ranging from enhancing workouts to detecting possible health issues [31]. These insights have led to enhanced user comprehension for companies, decreased turnover, and increased satisfaction. AI customizes each engagement, ranging from gamified tasks to online mentorship. Companies such as Peloton employ AI to personalize class suggestions and create engaging user experiences. This customization has been associated with a 15% increase in user satisfaction and retention, underscoring the significance of personalized experiences in digital platforms. Integrating AI into the fitness sector allows the business to be a dedicated supporter of its users [32]. AI streamlines repetitive tasks such as appointment setting, busy hour monitoring, and support services. Predictive analytics enhance staffing and resource distribution at gyms and fitness centers. These developments demonstrate how AI fitness innovations assist companies in redirecting their attention from routine operations to long-term expansion and growth. AI has revolutionized home fitness platforms such as Tempo and Mirror by offering users interactive, immediate feedback while they work out. These provide a gym-inspired environment that allows individuals to train efficiently from any location. To meet the rising demand for virtual fitness, companies need to integrate AI fitness application solutions. AI in fitness has transitioned from a luxury to an essential competitive advantage. By personalizing experiences and enhancing productivity, AI enables fitness companies to surpass their rivals. With the rapid advancement of AI in the fitness sector, businesses that adopt this technology are more likely to provide outstanding user experiences and achieve sustainable growth [33].

4.3.1 AI in Fitness with the Future of Personalized Workouts and Nutrition Plans

Artificial intelligence (AI) is transforming numerous sectors, and the fitness industry is included. As AI technologies advance, the future of customized workouts and nutrition plans offers a more personalized and effective method for health and fitness [34].

This change is poised to greatly benefit consumers, offering them tailored advice while also affecting fitness experts in multiple ways. Nonetheless, the trustworthiness of these AI applications is still an important issue that requires thorough attention. AI in fitness can generate extremely customized workout routines by using extensive data and advanced algorithms. Here, an in-depth look at the kinds of custom workouts that AI can generate is presented. AI programs can modify exercise routines instantly on the basis of an individual's performance. For example, if a user continually has difficulty with a specific exercise, the AI can adjust the workup to lower the intensity or propose different exercises that focus on the same muscle group. On the other hand, if the user is performing well, AI can increase the difficulty of maintaining ongoing advancement. Regardless of whether the aim is to lose weight, build muscle, enhance endurance, or increase flexibility, AI can create workout routines that accurately match these goals.

AI can also play a key role in preventing injury and supporting rehabilitation. Through the use of motion capture technology and wearable sensors to analyse movement patterns, AI can identify poor forms and offer immediate feedback for corrections. Moreover, with the oversight of a healthcare expert, AI can generate customized rehabilitation exercises for people to heal from injuries, making sure that they adhere to a secure and efficient recovery process. To address the boredom that often arises from repetitive workout routines, AI can introduce a diverse range of exercises to maintain user interest [35, 36].

4.4 Case Studies: Successful Workplace Wellness Implementations

I. Google

Naturally, individuals anticipate that Google would offer an excellent corporate wellness program given their reputation for impressive campuses and being a highly sought-after workplace [37]. Google provides a wide array of distinctive employee wellness benefits. Employees have the option to swim in the campus lap pool, engage in ping pong, join fitness classes such as kickboxing, or relax in a nap pod. Google's strategy for well-being is quite comprehensive. They motivate employees to instruct classes for fellow Google employees, enabling them to leverage their strengths. These courses encompass fitness classes, along with numerous other topics.

II. Motley Fool

The investment advisory firm—Motley Fool—also urges its staff to participate in numerous corporate wellness initiatives. In the Mashable article, Chief Wellness Fool Sam Whiteside describes that every month, they release a newsletter featuring a peer-nominated employee who is thriving and motivating others in the wellness area. The newsletter is named The Flex. Whiteside also creates a new fitness challenge each month. For example, a task in which employees were required to ensure that one meeting each day was engaging. From pushup pauses during a meeting to discussions on the move, staff continually speculated on what would motivate them to rise from their chairs, while those conducting meetings could employ their inventiveness to maintain employee involvement [38]. The motley fool promotes wellness by providing 50% reimbursement for race entry fees. This benefit encourages employees to participate in races as a group and offers an opportunity for individuals from various departments to connect outside of work hours. Whiteside also assists employees in selecting races to participate in and provides training support for those who are new to running. Motley Fool's engagement in corporate wellness stands at 86%.

III. Zappos

Zappos excels in delivering shoes and other products to its customers, but it also leads the way in corporate wellness initiatives. They provide typical corporate wellness benefits such as gym memberships, race reimbursements, and nap areas. However, they also desire individuals to choose to engage in physical activity naturally [39]. The wellness coordinator Kelly Maher views this as essential for the success of any corporate wellness initiative. Maher organizes groups of employees for various activities, such as golf lessons or visits to an indoor trampoline park—experiences they might be interested in but likely would not pursue alone.

IV. Draper, Inc.

Draper offers innovative solutions across various areas, primarily focusing on audio-visual and presentation products, window treatments, and solar management options, as well as gym and sports equipment. This business has existed for a long time—its founder, Luther O. Draper, started producing window shades in 1902! A key reason for their success throughout the years is their deep concern for their employees. Draper employees can maintain their fitness at the Wellness Park, which features an onsite track and workout stations. The Wellness Park features volleyball courts and table tennis. Workers are invited to take time to stroll around the track or initiate a game of casual volleyball to stay active during the day. This company in Indiana also faces challenges such as weight loss competition, where employees can earn cash prizes and gift cards to lose excess weight. The wellness coordinator, Linda Brinson, is passionate about wellness. She offers Weight Watcher sessions, Zumba classes, and additional options. While each of these organizations offers impressive corporate wellness programs for its staff, there are numerous insights one can gain and apply to one's own business. He might find himself in a situation within the business where he can develop something substantial, such as basketball courts or areas for relaxation and massage [40].

V. Johnson and Johnson

Given its history in healthcare, Johnson and Johnson is expected to emphasize corporate wellness and mental well-being. The organization's wellness initiative, "Live for Life," emphasizes comprehensive employee well-being, which includes (i) physical fitness, (ii) stress reduction, (iii) nutritious eating, and (iv) life-work equilibrium. Johnson and Johnson's enduring dedication to corporate wellness solutions has yielded measurable benefits. The company announced a return on investment of as much as $2.71 for each dollar invested in health programs [41]. By capitalizing on the wellness of their workforce, they appreciate enhanced efficiency and reduced healthcare expenses.

VI. Adobe

At Adobe, nurturing creativity is closely linked to advancing employee wellness initiatives. Adobe has created strong corporate wellness initiatives that emphasize the significance of mental well-being. Their efforts include (i) mindfulness sessions, (ii) adaptive work hours for improved work–life harmony, and (iii) an encouraging atmosphere fostering open conversations on mental well-being. Consequently, Adobe has established a dynamic workplace culture in which staff are motivated to express ideas openly while prioritizing their mental and physical wellness [42]. This has resulted in heightened innovation and reduced absenteeism in the organization.

VII. Microsoft

Microsoft recognizes that health and fitness in the workplace need to evolve to address the changing requirements of its employees worldwide. They have effectively launched a versatile wellness program that encompasses (i) options for telecommuting, (ii) availability of health apps, and (iii) workforce support programs (WSPs) [43, 44]. These regulations guarantee that workers uphold their work–life equilibrium, which is essential for ongoing motivation and efficiency. Microsoft's focus on corporate fitness and mental well-being has increased employee morale and enriched the overall company culture, highlighting the significance of flexibility in wellness initiatives.

4.5 Conclusions

Artificial intelligence (AI) is revolutionizing workplace wellness initiatives, providing organizations with creative methods to enhance employee health, involvement, and efficiency. Leveraging data and customized insights, AI-powered solutions can detect health patterns, suggest specific interventions, and promote a culture of wellness. These initiatives can include mental health chatbots driven by AI, fitness monitoring, and predictive analytics for proactive intervention. Although AI improves efficiency and personalization, it also brings significant issues related to privacy, data security, and ethical usage. Organizations need to address these issues openly to foster trust and guarantee responsible execution. AI can greatly enhance workplace wellness programs by making them more proactive, flexible, and effective. With the progression of technology, finding a careful equilibrium between innovation and ethical accountability will be essential for realizing the complete advantages of AI in fostering healthier, more fulfilling workplaces.

References

1. Mittal, P., Jora, R. B., Sodhi, K. K., & Saxena, P. (2023). A review of the role of artificial intelligence in employee engagement. In *Proceedings of the 2023 9th international conference on advanced computing and communication systems (ICACCS)* (Vol. 1, pp. 2502–2506). IEEE.
2. Priya, M. S., Khan, S., Siddiqui, S., Sharma, M. S., & Verma, S. (2024). The role of AI in shaping the future of employee engagement: Insights from human resource management. *Library of Progress-Library Science, Information Technology and Computer, 44*(3), 452.
3. Mer, A., & Srivastava, A. (2023). Employee engagement in the new normal: Artificial intelligence as a buzzword or a game changer? In *The adoption and effect of artificial intelligence on human resources management, Part A* (pp. 15–46). Emerald Publishing Limited.
4. Rao, S., Chitranshi, J., & Punjabi, N. (2020). Role of artificial intelligence in employee engagement and retention. *Journal of Applied Management-Jidnyasa, 15*, 42–60.
5. Dutta, D., Mishra, S. K., & Tyagi, D. (2023). Augmented employee voice and employee engagement using artificial intelligence-enabled chatbots: A field study. *The International Journal of Human Resource Management, 34*(12), 2451–2480.
6. Gusti, M. A., Satrianto, A., Juniardi, E., & Fitra, H. (2024). Artificial intelligence for employee engagement and productivity. *Problems and Perspectives in Management, 22*(3), 174.

7. Ravichandran, N., Inaganti, A. C., & Muppalaneni, R. (2023). AI-driven sentiment analysis for employee engagement and retention. *Journal of Computing Innovations and Applications, 1*(01), 1–9.
8. Alzeiby, E. A., Islam, N., Shaik, A. S., & Yaqub, M. Z. (2025). AI adoption in enterprises for enhanced strategic human resource management practices: Benefiting the employee engagement and experience. *Journal of Enterprise Information Management, 15*, 1334.
9. Saxena, M., & Mishra, D. K. (2025). Artificial intelligence: The way ahead for employee engagement in corporate India. *Global Knowledge, Memory and Communication, 74*(1/2), 111–127.
10. Gao, X., & Feng, H. (2023). AI-driven productivity gains: Artificial intelligence and firm productivity. *Sustainability, 15*(11), 8934.
11. Seamans, R., & Raj, M. (2018). *AI, labor, productivity and the need for firm-level data* (No. w24239). National Bureau of Economic Research.
12. Al Naqbi, H., Bahroun, Z., & Ahmed, V. (2024). Enhancing work productivity through generative artificial intelligence: A comprehensive literature review. *Sustainability, 16*(3), 1166.
13. Damioli, G., Van Roy, V., & Vertesy, D. (2021). The impact of artificial intelligence on labor productivity. *Eurasian Business Review, 11*(1), 1–25.
14. Li, E. Y., & Jan, A. (2023). Impact of artificial intelligence (AI) in enhancing productivity and reducing stress among students.
15. Yang, C. H. (2022). How artificial intelligence technology affects productivity and employment: Firm-level evidence from Taiwan. *Research Policy, 51*(6), 104536.
16. Alderucci, D., Branstetter, L., Hovy, E., Runge, A., & Zolas, N. (2020). Quantifying the impact of AI on productivity and labor demand: Evidence from US census microdata. In *Allied social science associations—ASSA 2020 annual meeting*.
17. Yi, Z., & Ayangbah, S. (2024). The impact of AI innovation management on organizational productivity and economic growth: An analytical study. *International Journal of Business Management and Economic Review, 7*(3), 61–84.
18. Edelman, B. G., Ngwe, D., & Peng, S. (2023). Measuring the impact of AI on information worker productivity.
19. Knight, B., Mitrofanov, D., & Netessine, S. (2024). The impact of AI technology on the productivity of gig economy workers. In *Proceedings of the 25th ACM conference on economics and computation* (pp. 833–833).
20. Yadav, S. J. (2025). Artificial intelligence and stress management tools for better mental health. In *AI Technologies and advancements for psychological well-being and healthcare* (pp. 163–186). IGI Global.
21. Dedeepya, R. R., Koneru, L. C., Lakshmi, A., & Kumar, G. K. (2022). Stress management system. In *Proceedings of the 2022 IEEE Delhi section conference (DELCON)* (pp. 1–5). IEEE.
22. Wang, R., Siau, K., & Zhang, Z. (2025). AI-assisted stress management: Efficacy compared to traditional methods.
23. Putri, F. R., Lestari, E., Hastiningsih, W. T., Yuberta, F., & Shomad, Z. A. (2025). Optimization of AI in monitoring and stress management for workers in the digital era. *International Journal of Community Care of Humanity (IJCCH), 3*(6), 1475.
24. D'Alfonso, S. (2020). AI in mental health. *Current Opinion in Psychology, 36*, 112–117.
25. Cross, S., Bell, I., Nicholas, J., Valentine, L., Mangelsdorf, S., Baker, S., et al. (2024). Use of AI in mental health care: Community and mental health professionals survey. *JMIR Mental Health, 11*(1), e60589.
26. Tutun, S., Johnson, M. E., Ahmed, A., Albizri, A., Irgil, S., Yesilkaya, I., et al. (2023). An AI-based decision support system for predicting mental health disorders. *Information Systems Frontiers, 25*(3), 1261–1276.
27. Carr, S. (2020). 'AI gone mental': Engagement and ethics in data-driven technology for mental health. *Journal of Mental Health, 29*(2), 125–130.
28. Luxton, D. D. (Ed.). (2015). Artificial intelligence in behavioral and mental health care.

29. Chen, I. Y., Szolovits, P., & Ghassemi, M. (2019). Can AI help reduce disparities in general medical and mental health care? *AMA Journal of Ethics, 21*(2), 167–179.
30. Talati, D. (2023). Artificial intelligence (AI) in mental health diagnosis and treatment. *Journal of Knowledge Learning and Science Technology, 2*(3), 251–253.
31. Lovejoy, C. A. (2019). Technology and mental health: The role of artificial intelligence. *European Psychiatry, 55*, 1–3.
32. Thakkar, A., Gupta, A., & De Sousa, A. (2024). Artificial intelligence in positive mental health: A narrative review. *Frontiers in Digital Health, 6*, 1280235.
33. Li, H., Zhang, R., Lee, Y. C., Kraut, R. E., & Mohr, D. C. (2023). Systematic review and meta-analysis of AI-based conversational agents for promoting mental health and well-being. *NPJ Digital Medicine, 6*(1), 236.
34. Dekker, I., De Jong, E. M., Schippers, M. C., De Bruijn-Smolders, M., Alexiou, A., & Giesbers, B. (2020). Optimizing students' mental health and academic performance: AI-enhanced life crafting. *Frontiers in Psychology, 11*, 1063.
35. Williams, D., Hodge, V. J., & Wu, C. Y. (2020). On the use of AI for generation of functional music to improve mental health. *Frontiers in Artificial Intelligence, 3*, 497864.
36. De Freitas, J., Uğuralp, A. K., Oğuz-Uğuralp, Z., & Puntoni, S. (2024). Chatbots and mental health: Insights into the safety of generative AI. *Journal of Consumer Psychology, 34*(3), 481–491.
37. Mody, V., & Mody, V. (2019, March). Mental health monitoring system using artificial intelligence: A review. In *Proceedings of the 2019 IEEE 5th international conference for convergence in technology (I2CT)* (pp. 1–6). IEEE.
38. Denecke, K., Abd-Alrazaq, A., & Househ, M. (2021). Artificial intelligence for chatbots in mental health: Opportunities and challenges. *Multiple Perspectives on Artificial Intelligence in Healthcare: Opportunities and Challenges, 15*, 115–128.
39. Olawade, D. B., Wada, O. Z., Odetayo, A., David-Olawade, A. C., Asaolu, F., & Eberhardt, J. (2024). Enhancing mental health with Artificial Intelligence: Current trends and future prospects. *Journal of Medicine, Surgery, and Public Health, 3*, 100099.
40. Umanandhini, D., & Karthika, S. (2024). Enabling mental wellness through interactive platforms. In *Proceedings of the 2024 9th international conference on communication and electronics systems (ICCES)* (pp. 1970–1975). IEEE.
41. Reyes-Portillo, J. A., Therefore, A., McAlister, K., Nicodemus, C., Golden, A., Jacobson, C., & Huberty, J. (2025). Generative AI—powered mental wellness Chatbot for college student mental wellness: Open trial. *JMIR Formative Research, 9*(1), e71923.
42. Singh, L. (2024). *Revolutionizing care: Developing an AI-human synergy mental health application for mental health worker's well-being* (Doctoral dissertation, Capella University).
43. Bhushan, B., Khanday, A., Aurangzeb, K., Sharma, S. K., & Nand, P. (Eds.). (2025). *Wellness management powered by AI technologies.* Wiley.
44. Wu, S. L., & Amini, M. (2025). Emotionally intelligent AI: Redefining wellness in the digital age. In *Navigating health information in the age of artificial intelligence* (pp. 37–66). IGI Global Scientific Publishing.

Chapter 5
AI Applications for General Well-Being

Artificial intelligence has surfaced as a potent instrument for influencing contemporary methods of enhancing human well-being, going well beyond productivity and efficiency to meet the comprehensive needs of individuals and communities. Through the integration of sophisticated data analysis, machine learning, and smart automation, AI is currently utilized to enhance mental, physical, and social well-being in various ways. Through individualized health tracking and early illness identification, as well as AI-driven therapy chatbots and fitness apps, technology is establishing more accessible and customized routes for healthier lifestyles. AI fosters inclusive settings by assisting individuals with disabilities, improving public safety, and promoting community involvement [1]. Additionally, its capacity to offer real-time insights and predictive solutions ensures proactive care rather than reactive measures. With well-being increasingly including mental strength, physical energy, and social ties, AI tools are becoming essential in fostering a balanced, supportive, and healthier community for everyone.

5.1 Defining Holistic Wellness Through Physical, Mental, and Emotional Health

Holistic wellness encompasses a person's total health by acknowledging the interrelationship of their physical, mental, and emotional well-being, understanding that a disruption in one aspect can affect the others. It encompasses a holistic approach to wellness, extending beyond merely addressing symptoms to consider the entire person, encouraging a sense of empowerment, and facilitating a balanced, satisfying lifestyle.

Wellness is a contemporary term with historical origins. The fundamental principles of wellness as both holistic and preventive can be traced to ancient cultures

S. Majumder and N. Dey, *AI-Driven Wellbeing to Enhance Lives and Work Environments*, Studies in Computational Intelligence 1257,
https://doi.org/10.1007/978-3-032-15735-5_5

from East China (India, China) to West China (Greece, Rome). During the nineteenth century in Europe and the United States, numerous intellectual, religious, and medical movements emerged alongside traditional medicine. These movements have established a solid basis for wellness today through their emphasis on holistic and natural methods, self-healing, and preventive care. Wellness-oriented and holistic approaches have become more prominent since the 1960s/1970s owing to the writings and insights of an informal group of US physicians and intellectuals (including Halbert Dunn, Jack Travis, Don Ardell, and Bill Hettler, among others) [2]. As these practices have advanced, spread, and become common, they have shaped the healthy living, self-help, self-care, fitness, nutrition, diet, and spiritual practices that form a thriving wellness movement in the twenty-first century. Holistic wellness is an all-encompassing approach to well-being that emphasizes the balance and unity of various aspects of health, including physical, mental, and emotional well-being. Physical health emphasizes sustaining the body via consistent exercise, balanced nutrition, sufficient rest, and preventive measures, guaranteeing that individuals possess the energy and vigor to live actively [3]. Mental well-being encompasses intellectual development, mental fortitude, and the ability to reason effectively, acquire knowledge, and adapt to new challenges. Emotional health revolves around acknowledging, understanding, and managing emotions effectively, promoting self-awareness, resilience, and healthy relationships.

Figure 5.1 depicts the six dimensions of wellness, indicating that complete well-being is achieved through a balance of various life aspects. At the core is Wellness, which is backed by six interconnected domains: physical wellness, which involves preserving health via exercise, nutrition, and active lifestyles; mental wellness, which focuses on enhancing the mind through learning and creativity; spiritual wellness, which prioritizes the discovery of purpose, significance, and inner tranquility; emotional wellness, which involves the effective recognition and management of emotions; social wellness, which underscores fostering positive relationships and community ties; and environmental wellness, which advocates living harmoniously with one's surroundings and nurturing the environment. Collectively, these aspects form a comprehensive strategy for attaining a healthy and satisfying life.

Fig. 5.1 Holistic wellness

5.2 The Role of Workplaces in Employee Well-Being and Managing Stress

Work environments significantly impact employee wellness by cultivating a nurturing culture, facilitating transparent communication, offering adaptable work options, promoting physical and mental health initiatives, and maintaining a healthy physical environment. By emphasizing these aspects, organizations can successfully decrease stress, enhance employee satisfaction and morale, and increase overall productivity and dedication, thereby fostering a flourishing workplace for both employees and the organization [4]. The environment of a workplace significantly influences employees' experiences. A positive work atmosphere can enhance employee well-being and reduce stress, whereas a harmful atmosphere may lead to fatigue and dissatisfaction. This study explores how workplace culture affects employee stress and well-being, offering guidance for fostering a healthier environment. Workplace culture encompasses the values, beliefs, behaviors, and practices that define an organization. A robust workplace culture aligns with the company's objectives and fosters a sense of belonging for employees. Effective communication serves as a foundation of a positive workplace culture. When workers believe that they can share their thoughts and worries, uncertainty and stress are diminished. Accessible communication fosters a feeling of safety, enabling employees to concentrate on their tasks without being distracted by concerns over misunderstandings or pending matters. On the other hand, insufficient communication can result in confusion and anxiety. When workers are uncertain about their responsibilities or what is expected of them, it can lead to a disordered atmosphere where stress flourishes [5].

Conversely, when efforts are overlooked, workers might sense a lack of appreciation, resulting in annoyance and tension. A culture that honors accomplishments, regardless of their size, can significantly improve employee well-being. A supportive workplace culture respects employees' personal lives and promotes a healthy work–life balance. When organizations encourage employees to take breaks and prioritize their health, it reduces the risk of burnout. Adaptable work arrangements, such as telecommuting options or flexible hours, allow employees to manage their responsibilities more efficiently, leading to reduced stress. Conversely, a culture that celebrates excessive work may result in ongoing stress and health issues. Workers might feel compelled to give up their personal time for work, leading to exhaustion and reduced efficiency [6]. Establishing robust support systems is essential for maintaining employee wellness. This encompasses access to mental health services, wellness initiatives, and employee assistance programs (EAPs). A work environment that prioritizes mental health strongly conveys that the well-being of employees is a top priority. When workers are aware that assistance is available, they are more likely to request help when necessary, thereby decreasing the likelihood of stress escalating into more significant problems [7]. Conversely, insufficient support can leave employees feeling isolated and stressed. Encourage an environment of openness where staff are at ease in expressing their opinions and issues. Regular team meetings, feedback discussions, and anonymous polls can help assess employee sentiment and address issues before they escalate. The most striking aspect is establishing recognition initiatives that honor employee accomplishments; recognizing effort through spoken compliments, prizes, or financial incentives can greatly increase morale and lessen stress. By urging staff to take breaks, utilize their vacation days, and unplug after work hours, adopting flexible work arrangements enables employees to balance their responsibilities more efficiently. Invests in wellness initiatives that emphasize mental well-being, physical fitness, and stress relief. Provide resources such as counselling services, fitness memberships, and workshops on techniques for reducing stress. Leaders need to exemplify healthy habits, emphasize employee wellness, and demonstrate their appreciation for work–life harmony. When leaders focus on their own health, it creates a beneficial model for the whole organization. Promote social engagement among staff via team-building activities and events. Fostering solid connections can improve teamwork and cultivate a feeling of belonging, lowering stress levels. Promoting employee well-being benefits both individuals and the organization. Encouraging well-being can aid in stress prevention and foster positive workplaces where people and organizations can flourish. Good health and wellness can serve as fundamental drivers of employee involvement and organizational effectiveness.

This factsheet concentrates on workplace wellbeing, detailing its significance. We describe the areas of our well-being model and examine the contributions of various stakeholders in fostering a healthy work environment. The impact of workplace culture on employee stress and wellness is significant. A supportive culture promotes transparent communication, appreciation, a balance between work and life, and assistance networks, all of which enhance employee satisfaction and resilience. Focusing on a healthy workplace culture allows organizations to lower stress levels and improve overall well-being, resulting in a more engaged and productive workforce [8].

5.3 Challenges in Achieving Holistic Wellness and Anxiety Management

The obstacles to maintaining comprehensive wellness and anxiety control frequently stem from the rapid pressures of contemporary life, as people find it difficult to balance their physical, mental, and emotional health. Insufficient time for self-care, unhealthy lifestyle choices, job-related stress, and a lack of understanding of holistic methods can obstruct overall well-being. Managing anxiety introduces an additional challenge, as ongoing exposure to stressors, excessive information, and pressure to perform can initiate or amplify anxious feelings and thoughts. Numerous individuals encounter difficulties in finding adequate support because of stigma, restricted access to mental health services, or unhelpful coping strategies. Tackling these challenges necessitates a proactive strategy that incorporates healthy habits, mindfulness, emotional control, expert advice, and nurturing social ties, enabling individuals to progressively attain comprehensive wellness while effectively handling anxiety.

Accomplishing comprehensive wellness and effectively handling anxiety requires a combined approach to physical, mental, emotional, and spiritual health [9]. Nonetheless, various obstacles may obstruct advancement. This is an in-depth examination of these difficulties, which is based on typical patterns observed in psychological, social, and health-related settings.

5.3.1 Complexity of Holistic Wellness

Comprehensive wellness involves harmonizing several facets of health—physical, mental, emotional, social, and spiritual. Every dimension possesses distinct requirements, and ignoring one can disturb overall wellness. For example, unhealthy eating, insufficient physical activity, or poor sleep can worsen anxiety and diminish resilience. Cognitive distortions or excessive thinking can complicate anxiety management [10]. Repressing feelings or having insufficient emotional control ability can result in increased stress. The absence of purpose or alignment with values can cause individuals to feel adrift, heightening anxiety. Managing efforts across these areas is labor intensive and demands self-awareness, discipline, and resource availability.

5.3.2 Prevalence of Anxiety Triggers

Anxiety is frequently caused by various external and internal influences that can be difficult to manage [11]. Job-related stress, financial strains, interpersonal conflicts, or worldwide uncertainties (such as economic instability or health emergencies) can lead to a persistent state of vigilance. In addition to these factors, negative thinking, perfectionist behavior, or unresolved trauma can continue cycles of anxiety, hindering individuals' ability to achieve relief. In the current digital era, an abundance of screen usage, comparisons on social media, and overwhelming information add to mental chaos and increased stress. Recognizing and reducing these triggers necessitates ongoing effort and perseverance, frequently requiring lifestyle modifications that can be difficult to maintain, such as decreasing screen time or tackling ingrained emotional problems.

5.3.3 Access to Resources

Holistic wellness typically requires access to resources such as nutritious food, secure exercise environments, mental health experts, or mindfulness initiatives. Nonetheless, therapy, fitness memberships, or wellness retreats can be costly [12]. Tight schedules hinder the ability to find time for activities such as meditation, journaling, or exercise. Rural or underserved regions might have inadequate access to mental health services or wellness communities.

5.3.4 Stigma Around Mental Health

Even with increasing awareness, stigma surrounding mental health remains prevalent in various cultures and communities. Worry about criticism frequently stops people from pursuing therapy or talking freely about their anxiety experiences. Numerous individuals might experience embarrassment about their challenges or view anxiety as a sign of personal failure, leading to hesitation in seeking help. In some cultures, stoicism is esteemed, or mental health problems are viewed as taboo, which further hinders open discussions [13]. Consequently, stigma may result in isolation, unresolved anxiety, and reluctance to engage in wellness activities that necessitate vulnerability, such as group therapy or mindfulness sessions.

5.3.5 Information Overload and Misinformation

The wellness sector is flooded with guidance, spanning from applications and social media personalities to self-improving literature and digital classes. Although some of these resources are supported by evidence, others do not have a scientific basis. Conflicting advice, such as "practice daily meditation" rather than "intense workouts are ideal," frequently leads to confusion for those looking for direction. Unverified wellness fads, such as rigid diets or untested supplements, may endanger both physical and mental well-being. Additionally, the excessive amount of information can hinder decision-making, complicating the process of pinpointing effective strategies. Navigating through false information demands critical analysis and significant time, which can be particularly daunting for individuals who are already facing anxiety.

5.3.6 Sustaining Motivation and Consistency

Achieving holistic wellness and managing anxiety necessitate a sustained commitment, yet staying motivated can be challenging. Techniques such as mindfulness or therapy might require weeks or months to reveal advantages, resulting in disappointment. Establishing new routines, such as consistent workouts or keeping a journal, demands self-discipline, which can be affected by stress or changes in life circumstances. Excessively dedicating oneself to wellness routines without equilibrium can result in fatigue, ironically heightening anxiety. Maintaining consistency during difficulties or gradual progress demands resilience, a trait that anxiety can weaken [14].

5.3.7 Individual Differences and Personalization

There is no universal method for wellness or managing anxiety: genetic factors, hormonal issues, or persistent health conditions can complicate anxiety control. Characteristics such as heightened sensitivity or a pursuit of perfection can intensify anxiety and make coping strategies more challenging. Distinct stressors, such as caregiving duties or employment uncertainty, require customized strategies. Discovering tailored strategies that cater to a person's specific requirements frequently requires experimentation, which can be frustrating and time-consuming.

5.3.8 Balancing Self-care with External Responsibilities

Numerous individuals balance various roles, such as parents, workers, partners, or caregivers, creating tension with self-care. Focusing on wellness activities frequently clashes with job deadlines or family responsibilities [15]. Allocating time for self-care may seem selfish, particularly in societies that prioritize productivity or nurturing others. High-performance workplaces might hinder taking breaks or days off for mental well-being. The incorporation of wellness into a hectic lifestyle demands the establishment of boundaries and effective time management, which can be challenging to cultivate during stressful periods.

5.3.9 Managing Expectations and Perfectionism

Anxiety frequently originates from or is aggravated by unattainable expectations: Pursuing perfect wellness habits (such as meditating flawlessly or eating "clean" all the time) can lead to extra stress. Social media and societal expectations can result in comparisons to others' seemingly flawless wellness paths. Concentrating on outcomes (e.g., "I need to eliminate anxiety") instead of the journey can result in discouragement. Adopting a mentality focused on progress instead of perfection necessitates cognitive restructuring, a challenge without proper guidance or practice.

5.3.10 Systemic and Environmental Factors

Wider systemic problems may hinder overall wellness and anxiety control: financial pressures can worsen anxiety and restrict access to wellness support. Insufficient community support or significant relationships can negatively impact mental health. City congestion, sound pollution, or insufficient green areas can impede relaxation and exercise. These elements are frequently out of personal influence, necessitating systemic alterations or inventive solutions that may not be readily achievable [16]. Education and awareness involve understanding evidence-based methods (such as cognitive-behavioral therapy and mindfulness) to manage misinformation. This process begins with simple routines, such as 5-min meditation sessions or brief walks, to establish consistency without feeling overwhelmed. Look for support groups, online communities, or reliable friends to lessen feelings of isolation and stigma. Trapists or coaches should be sought for tailored strategies, even if just periodically, because of expenses. Advocate for employee wellness initiatives or local resources to address systemic obstacles. Cultivate self-compassion to combat perfectionism and feelings of guilt, emphasizing advancement instead of flawlessness—Challenges. Several obstacles hinder our comprehension of effective workplace interventions aimed at improving employee well-being, as well as our grasp of workplace well-being as a whole. According to the 'call for papers', the evolving work environment presents various ongoing and novel challenges for HR professionals to handle, as well as for the overall sustainability of well-being in the existing and future workforce. A significant example involves swiftly evolving technologies that impact our work methods, the jobs we perform, and the ways we communicate and interact with coworkers. In addition to enduring risk factors such as heavy workloads, new challenges, including toxic leadership, bullying, and cyberbullying, have arisen. These have grown partly because of heightened pressure on organizations to surpass rivals and to remain viable amid these challenging economic conditions. This is becoming more crucial for managing well-being, yet it contrasts with a setting of limited resources and diminished security and stability in work environments. As previously mentioned, assessing well-being presents its own difficulties, as it becomes challenging to compare studies when the metrics employed differ and when definitions of well-being vary across research. According to the OECD [17], "Well-being inherently comprises multiple dimensions, making it challenging to summarize," which affects measurement approaches. Although current theories concerning well-being recognize the importance of resources, limited attention has been given to the expected availability of these resources and their duration [18]. A division in the literature also persists regarding the sector and the variations in the size of organizations. For example, SMEs often receive less attention, whereas larger companies might be overrepresented; however, they may also have greater resources to seek out and invest in initiatives. Scholars have noted an increasing focus on studies concerning SMEs. These issues provide further reasons why a 'one size fits all' approach may not effectively enhance employee well-being. Recent advancements in the HRM literature emphasize the importance of various components of the implementation process for interventions (including who, when, how, and to what extent) in attaining enhanced individual and organizational outcomes and are therefore embraced and encouraged. The uncertainties surrounding causality attribution and the lack of longitudinal studies lead to limited cost-benefit assessments of well-being programs. The evolving workplace has continuous effects on HR as a field, affecting its roles and functions, as well as well-being, owing to shifting working conditions and the pressure of achieving more with fewer resources, as challenges related to work intensification remain common. Emphasizing personal traits and coping strategies, such as mindfulness and resilience training, is highly beneficial for enhancing individual psychological capital but may unintentionally divert attention from what organizations can contribute to fair work, equitable pay, working conditions, or necessary legislative reforms [19]. Consequently, it is appropriate to highlight a special issue that emphasizes well-being and HRM in the evolving work environment. The seminar series and this special issue, which emphasize interdisciplinarity, highlight the benefits of adopting a more comprehensive perspective on well-being and HRM in the evolving workplace.

5.4 AI Applications for Mental Health

5.4.1 *Early Detection and Diagnosis*

AI leverages machine learning (ML), deep learning (DL), and natural language processing (NLP) to examine various data sources, including electronic health records (EHRs), speech patterns, social media interactions, or biometric data, to detect early indications of mental health disorders such as depression, anxiety, or suicidal thoughts. Vanderbilt University Medical Center created an ML algorithm that uses hospital admission information (such as age, sex, and previous diagnoses) to assess suicide risk with 80% accuracy. The University of Florida is evaluating an AI system for the early diagnosis of Parkinson's disease, which frequently presents with mental health symptoms.

5.4.2 *Personalized Treatment Plans*

AI algorithms analyse personal data (genetics, health history, lifestyle, and behavior patterns) to suggest customized interventions, including particular treatments or medications. AI models analyse genetic profiles to predict reactions to antidepressants, minimizing trial-and-error in medication management. Reinforcement learning (RL) tailor virtual reality (VR) exposure therapy according to patient responses, enhancing therapeutic results. It increases treatment effectiveness by personalizing interventions to meet individual requirements, leading to better patient outcomes.

5.4.3 AI-Powered Chatbots and Virtual Therapists

Conversational AI, which frequently leverages NLP, offers real-time mental health assistance via chatbots or virtual assistants that offer cognitive–behavioral therapy (CBT), mindfulness practices, or crisis management. Wysa: An AI chatbot employing CBT methods, mood monitoring, and guided meditation was demonstrated to enhance user-reported mental health issues. Woebot: Employs sentiment analysis to identify patterns of depression or hopelessness and recommends professional assistance when necessary. Limbic: An AI chatbot increased mental health service referrals by 15% in the UK's NHS, Talking Therapies, especially minority groups [20, 21].

5.4.4 Behavioral and Emotional Monitoring

AI examines data from wearables, mobile applications, or social platforms to track mood, sleep, or behavioral trends, offering insights into triggers for mental health. Biobeat: A device worn on the body that monitors sleep, activity, and heart rate variability to evaluate mood and cognitive condition, providing predictive alerts for timely intervention. Digital phenotyping uses smartphone information (e.g., typing pace, app interactions) to evaluate mental health [22].

5.4.5 Clinical Support and Administrative Efficiency

AI simplifies tasks such as updating medical records, consolidating patient information, or examining therapy session transcripts, enabling clinicians to concentrate on patient care. Lyssn, an AI application that streamlines therapy session notes and detects patterns for better therapist insights, aids psychiatrists by automating documentation, a labour-intensive process, and increasing diagnostic precision. This alleviates clinician burnout and enhances care quality by allowing more time for compassionate, person-centered interactions.

5.4.6 Mental Health Awareness and Education

AI-powered platforms examine social media or generate educational materials to increase mental health awareness, diminish stigma, and offer coping techniques. NLP tools analyse social media to assess public sentiment regarding mental health, guiding awareness campaigns [23]. Chatbots provide easily accessible self-help tools and link users to professional support. This enhances community awareness and promotes the tendency to seek assistance, particularly among young people.

5.4.7 *Immersive Therapies with Virtual Reality (VR)*

AI integrated with VR forms immersive therapeutic settings, such as eXtended-Reality Artificial Intelligence Ally (XAIA), which provides CBT through a trained virtual avatar. The XAIA program at Cedars-Sinai demonstrated favourable results in one study, with participants describing it as "empathetic" and "empowering" for addressing anxiety and depression. Interactive, adaptable therapy alternatives are offered, especially for individuals reluctant to pursue conventional treatment. AI-driven tools can examine speech patterns, facial expressions, text messages, and data from wearable devices to identify early indications of depression, anxiety, PTSD, or other conditions. This aids in prompt intervention before situations deteriorate. AI can customize therapy and treatment recommendations according to a person's specific symptoms, health history, and reaction patterns [24]. This customization enhances the effectiveness in contrast to generic methods. Chatbots and virtual counselors offer 24/7 assistance, coping techniques, mood monitoring, and crisis support when human experts are inaccessible. Certain people are reluctant to pursue therapy because of stigma, expense, or availability. There is a lack of mental health specialists globally. AI facilitates extensive support, connecting millions at once via applications, digital platforms, and wearable technology. AI can consistently monitor user actions and notify healthcare professionals if signs of relapse (e.g., depression or addiction) are identified, facilitating preventive care. AI-integrated smart devices can track physiological and behavioral metrics such as sleep quality, stress levels, and heart rate variability, which are closely associated with mental health.

5.5 AI Applications for Physical Health

AI algorithms evaluate X-rays, MRIs, CT scans, and ultrasounds to identify issues such as tumors, fractures, or heart disease with notable precision and quickness, frequently aiding physicians in early detection [25]. Through the analysis of health records, genetic information, and lifestyle habits, AI forecasts the likelihood of chronic diseases such as diabetes, heart disease, or cancer, assisting people in preventive actions. AI-powered applications and wearables monitor activity, sleep, and dietary habits; offer personalized fitness and nutrition plans to enhance overall physical well-being and performance; and are connected with IoT devices that track vital signs (blood pressure, glucose levels, and oxygen saturation) in real time, facilitating ongoing care for individuals with chronic conditions. Surgical robots enhanced by AI support physicians in conducting accurate and minimally invasive procedures, decreasing recovery periods and the likelihood of complications, accelerating the discovery of possible drug candidates, and forecasting their efficacy, resulting in swifter treatments for physical ailments. Chatbots and AI helpers offer reminders for medications, health appointments, and lifestyle tips, encouraging improved daily health management. AI-powered tools create customized rehabilitation plans for patients healing from injuries, strokes, or surgeries, monitoring their progress in real time.

Gadgets such as smartwatches employ AI to track heart rates, ECGs, oxygen levels, and activity trends, providing insights into physical well-being and identifying early warning indicators. Physical therapy often centers primarily on physical symptoms, ignoring other factors that affect recovery, such as mental health, nutrition, and lifestyle. Research indicates a connection between patients' negative emotions toward therapy and the emergence of depression, anxiety, fears, or anger related to their illness, impacting adherence [26, 27]. By embracing a comprehensive viewpoint, healing can be focused on in a way that takes into account all aspects of well-being—physical, mental, and emotional. Thus, complete recovery is assured, and ongoing well-being is secured rather than just temporary alleviation. AI-driven physical therapy enhances both the delivery and experience of therapy through the use of artificial intelligence (AI). Rehabilitation improves through the combination of therapists' expertise and advanced technology, resulting in a more customized, efficient, and accessible approach.

In essence, AI-driven physical therapy consists of employing smart systems to assess data, monitoring advancements, and offering personalized suggestions. These systems can understand movements, forecast recovery schedules, and even modify treatment plans dynamically according to progress, all while ensuring active participation in the process.

One of the most impactful applications of AI in physical therapy is the curation of highly personalized treatment plans (Fig. 5.2). Data from multiple sources, such as medical history, imaging results, and baseline performance, are analysed via AI-driven systems to tailor exercises to the needs of individuals. Through this level of individualization, each step of recovery is ensured to align with unique goals and physical conditions. In addition to customized plans being created, predictive analytics are used by AI to estimate recovery timelines. By analysing data from thousands of similar cases, insights are provided into how quickly mobility or strength might be regained. Realistic expectations and milestones are thereby set, keeping patients motivated throughout recovery. The way in which progress is monitored by patients and therapists is being revolutionized by wearable devices equipped with AI. Metrics such as joint angles, range of motion, and muscle activity during exercise are tracked by these devices. These data are interpreted by AI to provide real-time feedback on technique, ensuring correct movement and injury prevention. Moreover, gadgets such as smartwatches and fitness bands can collect different data, including heart rate, oxygen saturation, posture, and sleep habits. This information can be examined to provide tailored feedback and suggestions, which can improve workout efficiency. AI-driven virtual helpers and chatbots offer 24/7 assistance for individuals in physical therapy. These tools lead through activities, respond to inquiries, and provide prompts to maintain commitment to programs.

Fig. 5.2 AI applications in physical health

Beyond exercise guidance, chatbots can also help address common patient concerns, such as managing pain or understanding how a particular movement contributes to recovery. This ongoing support helps bridge the gap between in-person sessions, making therapy feel less isolating. AI is also being used to improve the diagnostic process in physical therapy. Therapists can use advanced tools to analyse patterns, postures, and ranges of motion precisely. For example, AI systems can detect subtle abnormalities in how people walk or move, which might indicate a movement disorder or potential complications.

These insights assist therapists in pinpointing issues more precisely and developing interventions that address the underlying causes of the condition. Timely diagnosis may also lower the likelihood of chronic problems, conserving time and resources for patients and healthcare providers. Research has indicated that stroke survivors utilizing an AI-enhanced robotic gait trainer, alongside goal setting, demonstrated greater advancements in confidence, balance, walking skills, and general quality of life than did individuals undergoing standard physical therapy [28]. AI plays a key role in enhancing the accessibility of physical therapy via telehealth services. Using AI tools, therapists can evaluate patients' movements from a distance, provide feedback, and modify their treatment plans without needing a clinic visit. Telehealth solutions powered by AI enable patients with mobility issues to access high-quality care without geographic limitations. These platforms utilize AI to automatically track progress, conserving time for both therapists and patients. AI-driven gamification makes physical therapy a more interactive experience. When exercise is transformed into engaging games, these systems maintain motivation and concentration on the rehabilitation objectives. Platforms such as XRHealth utilize extended reality (XR) settings that transform therapy into a more game-like experience. Research has indicated that utilizing the Nintendo Wii Fit for therapy aided knee osteoarthritis (OA) patients in enhancing muscle strength, physical performance, and balance while lowering their fall risk. The initiative also resulted in high patient involvement, with a retention rate of 83% and an adherence rate of 78%, highlighting the possible advantages of incorporating interactive games in OA therapy.

Figure 5.3 illustrates the "Benefits of AI in Physical Therapy," emphasizing four main benefits that artificial intelligence offers to rehabilitation methods. This finding demonstrates that AI can enhance recovery results by customizing treatment plans and monitoring patient progress in real time. It also facilitates more effective care, minimizing tedious manual tasks for therapists and allowing them to concentrate more on patient needs. Moreover, AI enhances accessibility, assisting patients who cannot attend in-person sessions to obtain remote therapy and support. Finally, AI enhances the cost-effectiveness of physical therapy by reducing the frequency of clinic visits and optimizing resource usage, benefitting both patients and health-care professionals. AI-driven tools enhance quicker and more efficient recovery by customizing therapy to fit personal requirements. Information from medical records, activity patterns, and wearable devices is utilized to develop individualized treatment plans for each patient. The motion range, power, and advancements during treatment are evaluated, and tactics are modified on the fly. Numerous studies have indicated that AI-driven physical therapy is positively received and regarded as beneficial and motivating by approximately 60–90% of participants. For numerous patients, traditional physical therapy is frequently viewed as daunting or lengthy, especially when appointment availability is restricted or when home exercise instructions are vague. AI streamlines the process with structured activities and instant feedback.

AI tools such as virtual assistants can enhance the understanding of exercise by highlighting exercise performance and significance. Wearable devices help track progress and correct forms during home exercises, reducing uncertainty about proper execution. Ongoing in-person oversight is unnecessary, as this technology guarantees that recovery stays on course. AI can facilitate access to care for individuals who are located far from physical therapy clinics or have difficulty accommodating appointments in their schedules. AI-driven telehealth platforms allow therapists to be accessed remotely, assessments to be conducted, and personalized support to

Fig. 5.3 Benefits of AI in physical therapy

be obtained from the convenience of home. Various AI-powered platforms, such as XRHealth, provide engaging and interactive alternatives to physical therapy, immersing patients in virtual environments. Regular in-clinic appointments, typically necessary in traditional physical therapy, can rapidly accumulate in terms of both time and expenses. AI lowers expenses by facilitating more efficient home-based programs and decreasing the number of clinic appointments [29].

Therapy apps and devices powered by AI can be much less expensive than multiple clinic appointments while maintaining high quality. With increased commitment to home exercises, AI can accelerate recovery, shortening the overall length of therapy. Studies indicate that the reduced expense of AI-driven virtual physical therapy compared with traditional in-person sessions leads to average savings ranging from $313 to $519. The projected savings encompass reduced provider visits, advanced imaging, surgical procedures, pain medication, and expenses for emergency care [30, 31]. The incorporation of AI into physical therapy is transforming patient rehabilitation and enhancing personalization, accessibility, and recovery efficiency. It enables patients to engage actively in their recovery process. As a patient, adopting AI-based physical therapy may result in quicker recovery, improved outcomes, and a more interactive experience.

5.6 AI Applications for Social Well-Being

In the constantly changing realm of technology, the combination of artificial intelligence with social consciousness has emerged as a strong catalyst, poised to revolutionize community development frameworks. Central to this transformation is the concept of "AI for Social Good," which transcends conventional perspectives of AI as merely a technological tool and frames it as a catalyst for meaningful social improvement. As an assistant professor of computer science, this research is crucial, as it links technological innovation with ethical obligations and the aim of shared improvement. Artificial intelligence (AI) has exceeded its former boundaries and now integrates into modern society, affecting multiple aspects of our daily lives. Progress in machine learning (ML) and artificial intelligence (AI) provides significant opportunities for tackling worldwide issues and enhancing social well-being in accordance with the United Nations' Sustainable Development Goals (SDGs). The AI for Social Goods (AI4SG) initiative aims to promote interdisciplinary partnerships between AI researchers and specialists in various fields to address urgent social challenges. This offers principles for creating effective collaborations in AI4SG, presenting knowledge on current initiatives and prospective possibilities for AI usage in social benefit activities. The importance of "AI for the social good" extends beyond theoretical ideas to practical projects and initiatives, including its application in healthcare. This tangible representation highlights our effort to influence real-world solutions outside the limits of academia [32]. In low-income nations, AI is seen as a possible means to reduce poverty, emphasizing its dual impact characteristics. With AI becoming a part of our lives, it is essential to proactively incorporate it into academic programs to prepare future generations with a comprehensive grasp of its consequences. In conclusion, the convergence of AI and social benefits calls for careful investigation and ethical involvement. While exploring this intricate discussion, our dedication to genuineness and precision remains unwavering, utilizing knowledge from credible sources and respected journals to enhance our comprehension. At this point, our educational group strives to engage significantly in discussions about responsible and effective technology utilization for the improvement of communities, reflecting the principles of AI for social goods.

AI applications aimed at social well-being encompass mental health assistance via chatbots and platforms for therapy matching, public health programs that facilitate disease tracking and disaster prediction, enhancing education through tailored learning systems, and environmental oversight to forecast and address climate risks [25, 33]. Other uses include assisting communication for individuals with speech disabilities, helping social workers by automating processes and pinpointing vulnerable individuals, and creating inclusive AI models to better address various populations.

Artificial intelligence significantly contributes to improving social welfare by strengthening relationships, minimizing disparities, and encouraging inclusive societies. Communication platforms powered by AI facilitate overcoming language obstacles via real-time translation, allowing individuals from various backgrounds to connect and work together more efficiently, thus promoting cultural awareness and international collaboration. Social media networks and virtual communities employ AI-powered moderation systems to identify harmful or abusive material, contributing to the establishment of safer online environments that safeguard individuals' mental and emotional well-being [34]. Moreover, AI promotes social inclusion by helping individuals with disabilities via applications such as vocal recognition, text-to-speech, image recognition for those with visual impairments, and advanced mobility devices, guaranteeing equal chances for societal engagement. AI improves public safety and well-being on a larger scale by examining crime trends, overseeing traffic conditions, and assisting in disaster management efforts to protect communities. Additionally, data analytics powered by AI assist governments and nonprofit organizations in recognizing at-risk groups, distributing resources efficiently, and creating focused social initiatives that address challenges such as poverty, joblessness, and educational disparities [35]. Through enhancing accessibility, safety, fairness, and inclusivity, AI not only addresses social issues but also enables individuals and communities to flourish in a more interconnected, supportive, and just world.

5.7 Conclusions

The core of conclusions involves merging insights, encapsulating main findings, and offering a distinct resolution to a debate, analysis, or reasoning. A solid conclusion not only reviews the key concepts discussed but also emphasizes their wider significance, giving the reader a feeling of fulfilment and guidance. It frequently highlights the insights gained, real-world applications, and possible directions for future investigations. Conclusions function as a link between analysis and action by cohesively tying together the introduction, main content, and conclusion of a piece, making a memorable impression on the audience and highlighting the importance of the topic.

References

1. Li, H., Zhang, R., Lee, Y. C., Kraut, R. E., & Mohr, D. C. (2023). Systematic review and meta-analysis of AI-based conversational agents for promoting mental health and well-being. *NPJ Digital Medicine, 6*(1), 236.
2. García-Madurga, M. Á., Gil-Lacruz, A. I., Saz-Gil, I., & Gil-Lacruz, M. (2024). The role of artificial intelligence in improving workplace well-being: A systematic review. *Businesses, 4*(3), 389–410.
3. Ruiz-Vanoye, J. A., Fuentes-Penna, A., Barrera-Cámara, R. A., Díaz-Parra, O., Trejo-Macotela, F. R., Gómez-Pérez, L. J., et al. (2025). Artificial intelligence and human well-being: A review of applications and effects on life satisfaction through synthetic happiness. *International Journal of Combinatorial Optimization Problems and Informatics, 16*(1), 1785.
4. Vetrivel, S. C., Arun, V. P., Maheswari, R., & Saravanan, T. P. (2024). Applications of AI techniques in healthcare and wellbeing. In *Machine learning and generative AI in smart healthcare* (pp. 61–92). IGI Global.
5. Majumder, S., & Dey, N. (2022). Knowledge management in various sectors. In *AI-empowered knowledge management* (pp. 37–61). Springer.
6. Zhou, Z., Asghar, M. A., Nazir, D., Siddique, K., Shorfuzzaman, M., & Mehmood, R. M. (2023). An AI-empowered affect recognition model for healthcare and emotional well-being using physiological signals. *Cluster Computing, 26*(2), 1253–1266.
7. Rosca, C. M., & Stancu, A. (2024). Fusing machine learning and AI to create a framework for employee well-being in the era of industry 5.0. *Applied Sciences, 14*(23), 2076–3417.
8. Revell, G. (2024). Generative ai applications in the health and well-being domain: Virtual and robotic assistance and the need for niche language models (nlms). In *Applications of generative AI* (pp. 189–207). Springer.
9. Goldstein, S., & Kirk-Giannini, C. D. (2025). AI wellbeing. *Asian Journal of Philosophy, 4*(1), 25.
10. Dai, Y., Chai, C. S., Lin, P. Y., Jong, M. S. Y., Guo, Y., & Qin, J. (2020). Promoting students' well-being by developing their readiness for the artificial intelligence age. *Sustainability, 12*(16), 6597.
11. Kjell, O. N., Sikström, S., Kjell, K., & Schwartz, H. A. (2022). Natural language analysed with AI-based transformers predict traditional subjective well-being measures approaching the theoretical upper limits in accuracy. *Scientific Reports, 12*(1), 3918.
12. Niedbał, R., & Pytel-Kopczyńska, M. (2024). The prospect of AI using in shaping the employees' well-being. *Scientific Papers of Silesian University of Technology. Organization and Management, 196*, 345–356.
13. Havrda, M., & Rakova, B. (2020). Enhanced well-being assessment as basis for the practical implementation of ethical and rights-based normative principles for AI. In *Proceedings of the 2020 IEEE international conference on systems, man, and cybernetics (SMC)* (pp. 2754–2761). IEEE.
14. Naor, N., Frenkel, A., & Winsberg, M. (2022). Improving well-being with a mobile artificial intelligence-powered acceptance commitment therapy tool: Pragmatic retrospective study. *JMIR Formative Research, 6*(7), e36018.
15. Ai, L. (2021). Artificial intelligence system for college students' physical fitness and health management based on physical measurement big data. *Wireless Communications and Mobile Computing, 2021*(1), 4727340.

16. Alsobhi, M., Khan, F., Chevidikunnan, M. F., Basuodan, R., Shawli, L., & Neamatallah, Z. (2022). Physical therapists' knowledge and attitudes regarding artificial intelligence applications in health care and rehabilitation: Cross-sectional study. *Journal of Medical Internet Research, 24*(10), e39565.
17. OECD. (2015). How's Life? 2015: Measuring Well-being. OECD Publishing. https://doi.org/10.1787/how_life-2015-en
18. Kim, D. S., Rodriguez, F., & Ashley, E. A. (2025). AI and digital health: Personalizing physical activity to improve population health. *Circulation: Cardiovascular Quality and Outcomes, 32*, e012416.
19. Ramasamy, L. K., Khan, F., Shah, M., Prasad, B. V. V. S., Iwendi, C., & Biamba, C. (2022). Secure smart wearable computing through artificial intelligence-enabled internet of things and cyber-physical systems for health monitoring. *Sensors, 22*(3), 1076.
20. Bohr, A., & Memarzadeh, K. (2020). The rise of artificial intelligence in healthcare applications. In *Artificial intelligence in healthcare* (pp. 25–60). Academic Press.
21. Xu, H. (2025). AI-powered technologies in physical education: Impact on physical and psycho-emotional well-being of students. *The Journal of General Psychology, 121*, 1–22.
22. Majumder, S., & Dey, N. (2022). Knowledge management system. In *AI-empowered knowledge management* (pp. 63–84). Springer.
23. Maher, C. A., Davis, C. R., Curtis, R. G., Short, C. E., & Murphy, K. J. (2020). A physical activity and diet program delivered by artificially intelligent virtual health coach: Proof-of-concept study. *JMIR mHealth and uHealth, 8*(7), e17558.
24. Shawli, L., Alsobhi, M., Chevidikunnan, M. F., Rosewilliam, S., Basuodan, R., & Khan, F. (2024). Physical therapists' perceptions and attitudes towards artificial intelligence in healthcare and rehabilitation: A qualitative study. *Musculoskeletal Science and Practice, 73*, 103152.
25. Guo, Q., & Li, B. (2021). Role of AI physical education based on application of functional sports training. *Journal of Intelligent and Fuzzy Systems, 40*(2), 3337–3345.
26. Li, Y., Li, Z., Duan, Y., & Spulber, A. B. (2023). Physical artificial intelligence (PAI): The next-generation artificial intelligence. *Frontiers of Information Technology and Electronic Engineering, 24*(8), 1231–1238.
27. Sumner, J., Lim, H. W., Chong, L. S., Bundele, A., Mukhopadhyay, A., & Kayambu, G. (2023). Artificial intelligence in physical rehabilitation: A systematic review. *Artificial Intelligence in Medicine, 146*, 102693.
28. Rahayu, N. I., & Ismail, A. (2023). Trends in the use of artificial intelligence (AI) technology in increasing physical activity. *Indonesian Journal of Educational Research and Technology, 3*(3), 295–304.
29. Gabarron, E., Larbi, D., Rivera-Romero, O., & Denecke, K. (2024). Human factors in AI-driven digital solutions for increasing physical activity: Scoping review. *JMIR Human Factors, 11*, e55964.
30. Xu, H. (2025). AI-powered technologies in physical education: Impact on physical and psycho-emotional well-being of students. *The Journal of General Psychology, 34*, 1–22.
31. Walimbe, V., Chaurasia, P., & Malani, R. (2020). Integrating AI into physical rehabilitation: Current applications and implementation challenges.
32. Wei, L., & Wang, S. J. (2024). Motion tracking of daily living and physical activities in health care: Systematic review from designers' perspective. *JMIR mHealth and uHealth, 12*(1), e46282.
33. Malliou, V. J., Pafis, G., Katsikas, C., & Plakias, S. (2025). Integrating physical activity and artificial intelligence in burn rehabilitation: Muscle recovery and body image restoration. *Applied Sciences, 15*(15), 8323.
34. Al-Akayleh, F., Al-Remawi, M., & Agha, A. S. A. (2024). AI-driven physical rehabilitation strategies in postcancer care. In *Proceedings of the 2024 2nd international conference on cyber resilience (ICCR)* (pp. 1–6). IEEE.
35. Zsarnoczky-Dulhazi, F., Agod, S., Szarka, S., Tuza, K., & Kopper, B. (2023). AI based motion analysis software for sport and physical therapy assessment. *Revista Brasileira de Medicina do Esporte, 30*, e20220020.

Chapter 6
AI and Organizational Wellness Culture

AI is reshaping the culture of organizational wellness by providing tailored assistance, anticipatory insights, and efficient processes to improve employee health and productivity. This involves employing AI to identify initial indicators of stress or burnout, offering customized resources, and streamlining administrative duties connected to wellness initiatives. Nevertheless, ethical factors such as data privacy and bias are essential for effective and responsible AI deployment in the workplace, as they can automate tasks such as scheduling wellness appointments, sending reminders, and monitoring participation in wellness activities, alleviating the administrative load on HR and wellness teams. AI-driven tools can offer workers instant feedback on their health and wellness, motivating them to implement beneficial changes and maintain progress towards their objectives [1]. Through personalized assistance and optimizing wellness programs, AI can foster a more encouraging and positive workplace, enhancing employee satisfaction and morale. When workers feel cared for and well, they tend to be more engaged and productive in their jobs [2]. Emphasizing employee wellness can result in decreased absenteeism and lower turnover rates, leading to cost savings for the organization and enhancing overall stability. AI can enhance communication and cooperation by recognizing chances for cross-functional collaboration and optimizing workflows. A culture that welcomes AI and ongoing education is more prepared to adjust to evolving business requirements and technological progress.

6.1 Entrenching AI in a Wellness Strategy

AI is revolutionizing wellness approaches by providing tailored, data-informed solutions for improving employee health. AI can customize wellness programs, identify potential health problems early, and offer personalized assistance for mental and physical well-being. This encompasses stress control, dietary advice, and ergonomic

S. Majumder and N. Dey, *AI-Driven Wellbeing to Enhance Lives and Work Environments*, Studies in Computational Intelligence 1257,
https://doi.org/10.1007/978-3-032-15735-5_6

enhancements, all aiding in a healthier and more efficient workforce [3]. AI can constantly modify programs using real-time data and employee input, guaranteeing maximum efficiency. By recognizing individuals at risk, AI can enable prompt actions, such as suggesting stress relief methods or linking employees to mental health experts. AI algorithms evaluate data to determine the impact of wellness initiatives and pinpoint areas needing enhancement. Organizations need to protect the privacy and security of employee information while utilizing AI in wellness initiatives. AI algorithms may exhibit bias; therefore, organizations must recognize possible biases and implement measures to reduce them. The incorporation of artificial intelligence into a wellness strategy changes how organizations and individuals address health and well-being. Through the use of sophisticated data analytics, customized suggestions, and immediate monitoring, AI can detect trends in behavior, stress, sleep, and activity, providing personalized interventions that enhance individual results. In workplaces, AI-powered platforms can help identify burnout early, recommend preventive actions, and promote an environment of ongoing support. At the personal level, wellness apps driven by AI can adjust to the changing requirements of users, including nutritional guidance and mental health tools. Integrating AI into wellness programs increases engagement and effectiveness while fostering a proactive, data-informed strategy for realizing comprehensive health objectives [4, 5].

6.1.1 AI Can Enhance Corporate Wellness

AI can improve corporate wellness in all six areas—physical, mental and emotional, nutritional, occupational, financial, and social—while also offering administrative assistance [6].

I. **Early Detection**

AI can assess personal health indicators to predict illnesses, allowing crucial early diagnosis. For example, AI might track heart rate, body temperature, and sleep habits to detect possible health concerns. Early detection in wellness involves recognizing diseases or health issues at their onset, typically before symptoms become apparent. This forward-thinking strategy enables prompt action, which could result in more efficient treatment, better results, and an enhanced quality of life. Early detection involves recognizing health issues in their early phases, prior to the appearance of any noticeable symptoms. This anticipatory strategy can result in more successful therapies, better results, and may ultimately preserve lives [7, 8].

II. **AI Coaching**

AI can provide customized health coaching designed for each employee's individual needs and objectives. For example, it might assist an employee in recuperating from an injury by creating a secure and efficient exercise program to rebuild strength. AI coaching in wellness involves the use of artificial intelligence to deliver tailored advice and assistance for people aiming to enhance their health and overall well-being. These AI-driven platforms serve as virtual trainers, providing personalized advice, encouragement, and data-informed insights to assist users in reaching their wellness objectives [9]. AI coaches evaluate numerous data factors, comprising user preferences, fitness levels, dietary requirements, sleep habits, and even emotional conditions sourced from wearables, applications, and user feedback. Through the use of this information, AI has developed personalized strategies for exercise, diet, rest, and stress reduction.

III. **Real-Time Health Tracking and Holistic Monitoring**

AI effortlessly combines with wearables for immediate health tracking and comprehensive monitoring. For example, it can assess an employee's exercise routines and recovery periods and then suggest training plans to enhance fitness. Employers also gain from fitness tracking. A company experienced a 25% reduction in employee sick leave after introducing AI-based fitness monitoring. Real-time health monitoring and comprehensive wellness tracking involve the ongoing and thorough evaluation of a person's health condition through wearable devices and digital health systems, allowing for tailored and preventative care [10]. This method advances past conventional, sporadic evaluations to deliver instant insights into an individual's health, enabling early identification of possible problems and customized solutions.

IV. **AI Therapist Chatbot**

AI chatbots can act as virtual counselors, assisting employees in managing different challenges and creating tailored strategies for achievement. Moreover, AI wellness applications can examine the language used by employees in emails or wellness initiatives to detect subtle indicators of stress, subsequently offering a set of guided inquiries to assist individuals in recognizing imbalances and, if necessary, suggesting professional assistance.

An AI chatbot might outperform traditional therapy: 68% of individuals prefer discussing stress and anxiety with a robot over their manager, whereas 80% are receptive to AI-assisted therapy (Oracle). Additionally, individuals utilizing AI therapy chatbots experienced a 64% greater decrease in symptoms of depression [11].

V. **Individualized Guided Meditation**

AI can offer customized meditation sessions designed to meet each employee's unique requirements, whether for relaxation, concentration, or improved sleep. It can also assess mood, stress levels, and past sessions to adjust in real time for a more personalized and effective experience [12]. A breathing exercise is fundamentally a mindfulness practice that entails slow, deep breathing with the diaphragm muscle to fill the lungs. The aim is to reduce our breathing rate and absorb more oxygen. When certain individuals envision meditation, they imagine a person quietly seated with closed eyes. However, in truth, meditation includes various types. It may be silent, or it may be led by a soothing voice.

VI. **Real-Time Stress Reduction**

AI is capable of detecting indicators of stress and can take action immediately. For example, AI-driven tools for employee well-being might detect when an employee's tone in emails indicates frustration and then suggest a short guided meditation to assist them in resetting. AI in wellness might save lives. Research has indicated that an AI algorithm is able to forecast suicide attempts within a week with 92% accuracy and within 2 years with 85% accuracy [13]. To effectively alleviate stress, individuals should consider adding activities that enhance relaxation, mindfulness, and physical health. Participating in hobbies, enjoying nature, using relaxation methods, and working out can all help alleviate stress. Moreover, focusing on sleep, adhering to a nutritious diet, and fostering social relationships are essential for controlling stress levels.

VII. **Diet Analysis**

AI can assist workers in examining their eating habits to pinpoint positive practices and aspects that may need enhancement. For example, workers might share pictures of their meals with an AI-driven chatbot [14]. The chatbot might monitor their eating patterns and recognize strengths (such as consuming many leafy greens) and weaknesses (such as excessive intake of processed foods). This finding might suggest easy modifications for improved nutrition. Analysing diet is vital for overall health, as it enables people to comprehend their eating patterns, pinpoint nutritional deficiencies, and make educated decisions to enhance their well-being. It involves evaluating eating habits, nutrient consumption, and food quality to encourage a balanced and healthy way of living.

VIII. **Meal Plans and Schedules**

AI can generate personalized meal plans, serving as a digital nutritionist. This is especially beneficial for individuals with dietary limitations and sensitivities [15]. For example, an employee who recently discovered that their child has a food allergy might utilize AI to create meal plans safely while gathering more information about the allergy. Meal plans and schedules in wellness pertain to the systematic arrangement, timing, and structuring of meals to enhance an individual's overall physical, mental, and emotional health. It generally includes the following: (i) nutritional balance, i.e., ensuring that meals supply proper macronutrients (carbohydrates, proteins, and fats) and micronutrients (vitamins and minerals) for optimal well-being; (ii) meal timing, i.e., determining when to eat (e.g., breakfast, lunch, dinner, and snacks) to sustain consistent energy levels, increase digestion, and synchronize with the body's natural patterns; and (iii) position control, i.e., regulating the amount of food to avert overeating or insufficient nutrition [16].

IX. **Burnout Prevention**

AI can track employee health indicators and utilize predictive analytics to detect initial signs of burnout among employees. For example, AI can examine work hours, time taken to complete tasks, communication patterns, and sentiment, along with physiological indicators such as heart rate variability, to predict burnout and take action. Burnout is a condition of physical, emotional, and mental fatigue resulting from extended stress, frequently due to job or caregiving duties [17]. This may result in decreased productivity, skepticism, and feelings of isolation. The avoidance of burnout necessitates active approaches to maintain health and equilibrium. Strategies to prevent burnout are essential. Set boundaries involve creating distinct separations between work and personal life. Restrictions over time, refraining from checking work emails after hours, and conveying one's boundaries to coworkers or supervisors. Focus on self-care by including consistent exercise, sufficient sleep (7–9 h), and a nutritious diet in daily habits. Even brief pauses for stretching or mindfulness can rejuvenate all. Practicing stress management involves the use of relaxation methods such as deep breathing, meditation, or yoga. These factors can lower stress hormones and enhance mental clarity [18].

X. **Financial Goal Planning**

AI platforms aid in financial goal planning, assisting employees in establishing personalized and attainable objectives to increase savings, lower debt, and prepare for retirement.

It can also assist employees in making the most benefits packages. An AI-driven financial wellness platform might alert employees when they are overlooking an employer match for their retirement contributions. Financial goal setting is a systematic approach that assists individuals and organizations in defining, prioritizing, and reaching their financial aims. It starts by evaluating present financial status, determining immediate and future objectives—such as saving for a house, establishing an emergency fund, or preparing for retirement—and then developing a practical, actionable plan to achieve them. Efficient planning requires defining measurable goals, creating budgets, and tracking progress to remain aligned. By consistently evaluating and modifying plans in light of life changes or market shifts, individuals can remain flexible and strong. In the end, planning financial goals promotes a feeling of control and assurance, allowing individuals to make knowledgeable choices, prevent excessive debt, and ensure their financial security for the future.

6.2 AI for Measuring Organizational Health and Culture

AI is progressively utilized to evaluate and gauge organizational well-being and culture by examining employee data, communication trends, and performance indicators. These tools offer insights into aspects such as employee engagement, team unity, and overall wellness, allowing organizations to make data-informed decisions to enhance their culture and promote a flourishing workplace. AI for assessing organizational health and culture involves the use of artificial intelligence methods and technologies to analyse, track, and enhance an organization's overall wellness, performance setting, and cultural coherence [19]. It utilizes data sources such as employee input, communication trends, HR documentation, performance metrics, and collaboration analysis to recognize patterns, evaluate engagement levels, uncover risks (e.g., burnout or turnover), and gauge compliance with organizational principles. Through techniques such as sentiment analysis, machine learning, and predictive modelling, AI delivers immediate, data-informed insights that allow leaders to improve workplace culture, bolster employee experience, and maintain organizational well-being.

6.2.1 AI with Organizational Culture

Organizational culture defines the workplace atmosphere and its impact on an employee's mindset, behavior, and perception of work. It can greatly affect performance, how individuals interact, their dedication, and the organization's appeal to new talent. Organizational culture can be defined as a collection of common beliefs held by the members of an organization; these collective beliefs set the organization apart from others. Organizations exhibit shared behavioral patterns that employees utilize to reach goals, which are imparted to newcomers and symbolize the unspoken and abstract aspects of an organization [20]. Previous studies indicate that organizational culture has a notable effect on financial performance and is more impactful than organizational strategy and structure. While organizational culture has been extensively studied, it remains intricate, and no single definition is widely agreed upon. It is commonly described as "a set of common assumptions, values, and beliefs that is manifested in its practices and objectives and additionally aids its members in comprehending the organizational functions." Organizational culture influences the difficulties that organizations encounter when implementing new technologies. Therefore, the use of AI necessitates significant transformations in the business and organizational culture of companies to attain precise decision-making and enhance innovation and performance. Artificial intelligence has a lengthy history, and with the continuous rise in research, it is difficult to pinpoint a singular and comprehensive definition for AI. On a meta-level, McCarthy describes it as being "focused on approaches to attaining objectives in contexts where the available information possesses a specific complexity." The techniques that must be employed are connected to the issue posed by the scenario and are alike regardless of whether the problem solver is a human, a Martian, or a computer application. While this definition outlines the broad capabilities of AI applications, its realization within an organizational setting can be specified through functions such as machine learning, robotics, natural language processing, expert systems, or speech recognition [21]. For a company to utilize and take advantage of the outlined functions, it must develop the necessary capabilities to perform AI (and its functions) instead of concentrating on its technological attributes. Therefore, AI capabilities refer to a company's ability to choose, coordinate, and utilize its resources specific to AI.

6.2.2 Quantitative Metrics for Measuring Health and Culture

Organizational health signifies an organization's capacity to operate efficiently, adjust to change, and maintain high performance consistently over time. It includes employee involvement, welfare, retention, and overall efficiency. Organizational culture refers to the collection of common values, beliefs, and behaviors that influence how tasks are accomplished within an organization [22]. It encompasses elements such as trust, psychological safety, communication, leadership, and inclusivity.

I. **Natural Language Processing (NLP) and Sentiment Analysis**

NLP and sentiment analysis are essential in wellness programs, as they facilitate the examination of text data to gauge user sentiment and pinpoint areas needing enhancement. NLP methods assist in analysing different types of text, such as social media content, feedback surveys, and patient documents, to derive valuable insights regarding well-being. Sentiment analysis, a fundamental NLP application, subsequently classifies these insights as positive, negative, or neutral, enabling specific interventions and program modifications. NLP can examine text data, such as social media updates, to detect language trends and keywords that might suggest symptoms of depression, anxiety, or other mental health issues. NLP can assist in recognizing individuals who might be facing a decrease in their mental health and require support by observing shifts in sentiment within user-generated content [23]. NLP is capable of assessing shifts in user sentiment following the introduction of particular interventions or mental health initiatives, facilitating the evaluation of their impact.

II. **Machine Learning and Predictive Analytics**

Machine learning (ML) and predictive analytics are revolutionizing wellness initiatives by facilitating tailored interventions and proactive risk mitigation. These technologies examine extensive health data to pinpoint individuals prone to particular health concerns, enabling customized wellness initiatives and focused assistance [24]. Machine learning algorithms, such as random forests, are capable of examining different data points (activity levels, sleep habits, stress indicators, etc.) to forecast the probability of experiencing specific health problems such as burnout or disease progression. By utilizing these forecasts, customized wellness initiatives can be developed and implemented to meet particular needs and risk elements.

Figure 6.1 depicts the cycle of Predictive Analytics, outlining the essential phases in creating and utilizing predictive models. The detailed explanation of every step is as follows. Defining a project involves recognizing the issue or business query that predictive analytics will solve and precisely detail the objectives, scope, and criteria for success. Data collection encompasses the accumulation of pertinent past and present information from diverse sources. It is crucial at this stage to guarantee the quality, completeness, and relevance of the data to the issue. Data analysis entails examining and comprehending the gathered data through statistical methods and visualization tools. Recognizing trends, patterns, and connections will guide model development. Statistical methods involve the use of statistical methods to organize data for predictive modelling. Conduct hypothesis testing, correlation analysis, and

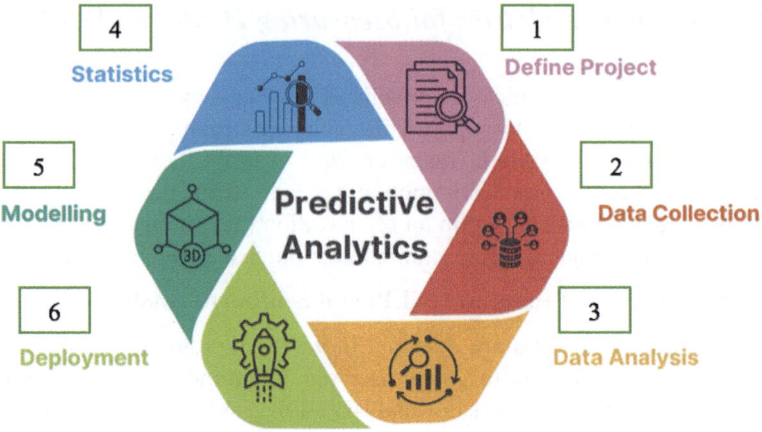

Fig. 6.1 Process cycle of predictive analytics

additional techniques to confirm the findings. Modelling consists of creating predictive models by using machine learning or statistical algorithms. Models for training and testing are necessary to guarantee precision and dependability. Deployment refers to the integration of the predictive model into actual systems or processes for decision-making. Track performance and adjust the model when necessary for shifting situations.

III. **Organizational Network Analysis (ONA)**

Organizational network analysis (ONA) can greatly improve workplace wellness initiatives by revealing social dynamics and communication trends that influence employee welfare. Through the mapping of these networks, organizations can recognize isolated individuals, identify communication bottlenecks, and grasp the dynamics of information flow, ultimately resulting in more focused and effective interventions.

Figure 6.2 depicts the organizational network analysis (ONA) model, which charts and examines the connections and communication pathways within an organization. It recognizes three essential kinds of network roles and their functions: (i) Central nodes: These nodes are the people who seem to know everyone. Key nodes rapidly disseminate extensive information and have considerable sway over groups. Central nodes can be located throughout the organizational structure, are often highly regarded, and exhibit a strong interest in industry updates and trends. (ii) Knowledge broker: These individuals act as connectors among different groups. In the absence of knowledge brokers, information and ideas cannot be exchanged. (iii) Peripheral: High-potential peripherals can endanger organizations, as they are frequently ignored and isolated from the rest of the company. For example, outstanding Java developers who fail to share their expertise in best practices with others hinder product development.

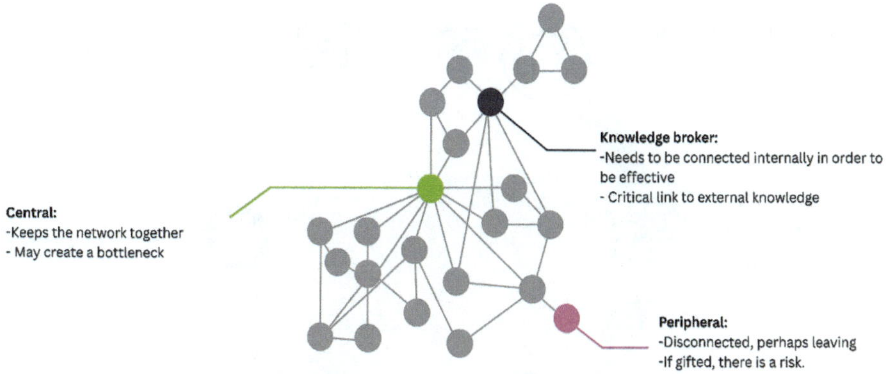

Fig. 6.2 The organizational network analysis (ONA) model

IV. **Computer Vision and Meeting Analytics**

Computer vision and meeting analytics can greatly improve wellness initiatives by offering insights into employee engagement and well-being. Computer vision can evaluate visual information from meetings to determine stress levels, activity levels, and social interactions, whereas meeting analytics can monitor attendance, the length of meetings, and communication trends. This information can be utilized to customize wellness programs, enhance participation, and detect possible health threats [25].

Computer vision systems undertake a sequence of actions to evaluate and interpret visual information (Fig. 6.3). This procedure starts with the acquisition of images or videos through cameras or various imaging tools. These visual data provide the basis for further examination. Unprocessed images go through initial processing stages, such as resizing, normalization, and noise minimization. During this phase, computer vision algorithms perform pattern recognition to identify significant features from the preprocessed images. Features that have been extracted are frequently expressed in a suitable mathematical format that can be handled by machine learning algorithms.

6.2.3 AI Tools and Platforms for Assessing Health in Organizations

AI tools and platforms for organizational health evaluation are digital resources that leverage artificial intelligence to assess, analyse, and enhance the overall wellness, efficiency, and culture of an organization. These platforms analyse data from various sources—including employee surveys, internal communications, performance metrics, wellness program participation, and behavioral trends—to deliver instant insights into elements such as engagement, mental well-being, productivity, inclusion, and workplace satisfaction.

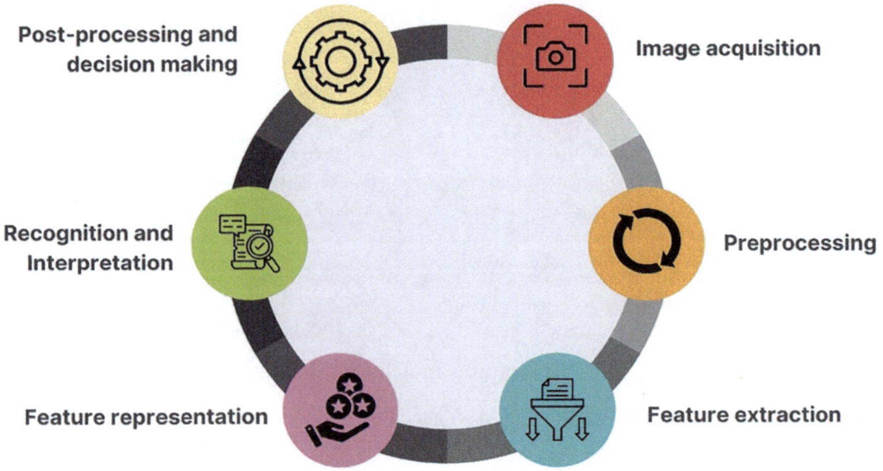

Fig. 6.3 Computer vision in healthcare

I. **Aiberry**

Aiberry, said “I-berry,” is a platform for mental health assessment that utilizes AI technology. It utilizes text, audio, and video inputs during short conversational assessments to provide immediate, measured risk scores and mental health information, such as signs of mood, focus, and energy levels. The AI assistant Botberry performs interactive evaluations by analysing conversation content, voice patterns, and facial cues to identify possible mental health issues such as depression or anxiety. HR departments can monitor employee mental health organization-wide in real time through anonymized trend dashboards. Aiberry seamlessly integrates with corporate wellness platforms, providing analytics that assist in tracking platform usage and wellness metrics over time [26]. Compared with conventional form-based evaluations, the chatbot format of Botberry is more engaging and tends to promote candid involvement. Assessment insights can inform focused wellness initiatives—enabling HR to meet particular needs and efficiently enhancing resource distribution.

II. **SAFER Peer Benchmarking Tool**

The SAFER Peer Benchmarking Tool is an analytical resource created by the Joint Commission to assist healthcare organizations in comparing their SAFER® (Survey Analysis for Evaluating Risk) Matrix performance against that of similar institutions. To improve health and safety practices within organizations, the tool uses collected survey data to highlight strengths, identify improvement areas, and compare performance with that of similar facilities. The SAFER Peer Benchmarking Tool allows leaders to visualize trends and identify compliance gaps in sectors such as leadership effectiveness, patient safety, and process reliability, helping them prioritize corrective measures and assess the impact of improvement efforts over time [27]. It fosters a culture of ongoing quality improvement by offering confidential, data-driven insights that promote peer learning and alignment with industry standards.

III. **Microsoft Viva Insights**

Microsoft Viva Insights is a component of the Microsoft Viva employee experience platform aimed at enhancing productivity, collaboration, and well-being at work through insights based on data. In the realm of health and wellness, Viva Insights functions as a "digital workplace wellness coach," assisting organizations in harmonizing employee performance with their mental, emotional, and physical health. Microsoft Viva Insights aims to assist individuals, managers, and leaders in comprehending work patterns and their effects on productivity and well-being. This utilizes aggregated and anonymized information from Microsoft 365 applications (Outlook, Teams, Calendar, etc.) to create insights [28]. This utilizes analytics and AI to convert workplace activity trends into tailored suggestions that enhance focus periods, work–life balance, and healthier work practices. Insights at the individual level remain confidential; leadership has access only to summarized and anonymized information.

IV. **Culture Amp, Glint (LinkedIn)**

Culture Amp and Glint (LinkedIn) are premier platforms for employee engagement and organizational wellness, aimed at assisting companies in assessing, comprehending, and enhancing workplace culture, performance, and well-being. Culture Amp emphasizes collecting comprehensive employee feedback via engagement, well-being, and inclusion surveys, alongside performance management tools and predictive analytics, to detect patterns such as burnout risk or turnover probability [28, 29]. Glint, connected with LinkedIn and Microsoft Viva, focuses on immediate pulse surveys, AI-powered insights, and manager dashboards that monitor sentiment, workload stress, and engagement factors, facilitating quick responses to enhance team morale and mental well-being. Both platforms serve as organizational "listening systems," transforming employee feedback into actionable strategies that boost culture, improve productivity, and encourage sustainable wellness. Table 6.1 illustrates how AI assesses organizational well-being and culture.

6.3 Encouraging Proactive Wellness Through AI Nudges

Artificial intelligence is subtly changing the standards of workplace wellness. No longer restricted to tracking steps or sending standard alerts, AI currently hears, understands, and reacts to the specific patterns of every employee's life. In essence, AI in corporate wellness involves the use of advanced technologies, including data-driven algorithms, chatbots, and predictive analytics, to develop wellness initiatives that are tailored to the individuals they assist.

The advantages are evident. Workers receive assistance tailored to their requirements and timetables, whereas companies experience increased participation,

Table 6.1 AI measures of organizational health and culture

AI capability	What it measures/analyses	Example tools/platforms
Sentiment analysis (NLP)	Spirit, tone of communication, involvement	Microsoft viva, worklytics
Predictive analytics	Turnover risk, burnout, engagement trends	HP predictive analytics
Network analysis (ONA)	Collaboration, silos, information flow	Humanyze, worklytics
AI-powered surveys	Engagement, psychological safety, feedback	Culture amp, glint, teamspective
Computer vision	Meeting engagement, participation equity	Meeting analytics tools
Multi-source integration	Holistic health and culture insights	Microsoft viva, glint
Bias detection	Equity in reviews, promotions, and communication	Custom AI tools

reduced absenteeism, and a stronger, more resilient workforce [30]. However, this transition presents new obstacles. In the background, a range of cutting-edge technologies is enhancing employee wellness to be more individualized, anticipatory, and efficient than ever before. From machine learning to natural language processing, these AI technologies quickly transform workplace wellness, providing customized solutions that cater to employees' physical and mental well-being and promote sustainable organizational growth.

I. **Machine Learning (ML)**

The initial topic is machine learning. It is central to most cutting-edge wellness platforms today. ML algorithms analyse vast amounts of data from wearable devices, health risk evaluations, and work behaviors to identify trends and forecast risks. This allows businesses to go beyond standard guidance and offer tailored fitness, nutrition, and stress management suggestions. Each can be customized to fit an employee's distinct profile and immediate requirements. Consider a system that detects when the stress increases and subsequently adjusts the wellness plan immediately. A 2023 SSRN research paper called "Impact of AI in Employee Wellness and Well-being Programs" emphasized that ML-based initiatives have resulted in "greater engagement and significant decreases in absenteeism and healthcare expenses" for companies that adopt them [31].

II. **Natural Language Processing (NLP)**

Furthermore, NLP enables AI to comprehend and reply to human language. This technology fuels chatbots and virtual mentors, which have become essential for employees in need of assistance.

Whether it is a late-night wellness check or a quick response to a benefits inquiry, NLP guarantees that help is available around the clock. Chatbots and Virtual Assistants offer conversational interfaces that provide mental health assistance, respond to wellness inquiries, and deliver coaching continuously, enhancing support accessibility and reducing stigma. In sentiment analysis, NLP examines text from employee surveys, emails, and chat logs to assess mood and grasp the emotional state of the organization. NLP streamlines intricate plan documents or health literature, simplifying information retrieval and assisting employees in making informed choices [32]. A systematic review in Nature Translational Psychiatry revealed that chatbots driven by NLP are transforming mental health assistance, providing discreet and scalable tools for managing stress, anxiety, and depression.

III. **Computer Vision**

Third, computer vision involves enabling AI to "see" and comprehend images or videos. In other words, it allows AI to analyse images and videos, creating new opportunities for workplace wellness. Research has indicated that "emotion recognition powered by AI can assist organizations in addressing mental health issues proactively before they intensify." This recognition will facilitate prompt assistance and compassionate wellness measures [33]. Although still developing, these tools are already aiding organizations in identifying concerns prior to them escalating into problems, promoting a safer, healthier workplace.

IV. **Predictive Analytics**

Then, predictive analytics uses both historical and real-time data to anticipate future health threats. AI can identify early indicators of burnout or chronic illness by analysing data from wearables, productivity applications, and digital communication habits. Research in IRE journals indicates that predictive analytics have enabled organizations to decrease employee sick days by as much as 25% and reduce burnout rates by 30%. For example, Deloitte employed predictive analytics to evaluate employee engagement and exhaustion. This enabled them to determine that 30% of their employees were inundated by digital interruptions [34].

V. **AI Agents and Chatbots**

Finally, AI-driven agents and chatbots have become essential for wellness assistance. These virtual assistants help employees navigate resources, provide personalized prompts, and assess mental health issues [35]. The effect is significant when these technologies collaborate. Workers receive personalized wellness suggestions suited to their specific requirements. Organizations can address minor issues before they escalate into major problems, and HR teams finally possess the resources to assess what is effective and what is not.

6.4 Conclusions

Artificial intelligence is swiftly changing how companies comprehend, cultivate, and maintain a culture of wellness in the workplace. By utilizing features such as sentiment analysis, predictive analytics, network mapping, and AI-driven surveys, organizations can transition from reactive wellness initiatives to proactive, data-informed strategies. AI allows leaders to identify early indicators of burnout, disengagement, and cultural discord while also offering insights to enhance inclusivity, equity, and psychological safety. When utilized ethically and with a keen emphasis on privacy, AI transforms from merely a measurement instrument into a driving force for cultural transformation that enables organizations to cultivate healthier, more resilient, and high-achieving workplaces. In the end, incorporating AI into wellness programs connects employee experience with organizational success, promoting a culture where well-being is a fundamental business priority rather than a secondary concern.

References

1. Ahlberg, J., & Eriksson, C. (2021). To measure organizational wellness with AI: A future competitive advantage?
2. Oladele, O. K. (2023). Data-driven work culture: How AI enhances employee well-being and reduces occupational stress.
3. McCall, A. (2023). From reactive to proactive: Using AI to redefine employee wellness in the tech industry.
4. Parasa, S. K. (2023). Impact of AI in employee wellness and well-being programs. *European Journal of Advances in Engineering and Technology, 10*(5), 98–100.
5. Pandey, A., Mishra, S., & Kadanga, A. (2025). Role of artificial intelligence in enhancing employee well. *Global Work Arrangements and Outsourcing in the Age of AI, 25*, 11542.
6. Rao, T. V. N., Patel, G. S., Bhavana, S., & Swetha, C. (2025). AI-driven decision support system for workplace wellbeing. In *Insights on wellbeing and happiness in the workplace* (pp. 1–18). IGI Global Scientific Publishing.
7. Jean, G. (2023). Leveraging centralized AI systems for proactive employee wellness management: Predictive modelling and work pattern optimization in the IT industry.
8. Rosca, C. M., & Stancu, A. (2024). Fusing machine learning and AI to create a framework for employee well-being in the era of industry 5.0. *Applied Sciences, 14*(23), 2076–3417.
9. Mary, B. J. (2015). *Predictive analytics for employee wellness: An AI-driven solution to minimize workplace burnout*. Britney Johnson Mary New York University United States.
10. Oluwagbade, E. (2025). Ethical AI in employee wellness: Balancing data-driven insights with employee privacy and trust.
11. Rubim, F., Rubim, L., & Thornton, A. R. (2020). Creating a culture of wellness. In *Power* (pp. 171–187). Academic Press.
12. Subramanyam, M. A. (2020) AI-driven wellness interventions for unlocking the potential of employee engagement in the era of industrial 4.0. In *Of phnom penh* (p. 35).
13. Sadeghi, S. (2024). Employee well-being in the age of AI: Perceptions, concerns, behaviors, and outcomes. arXiv preprint arXiv:2412.04796.
14. Putrevu, D. J., & Mertzanis, C. (2025). The rise of the wellness business sector: A systematic literature review.

15. Kuru, H. (2023). Understanding employee wellness in industry 5.0: A systematic review. *Journal of Ekonomi, 5*(1), 32–35.
16. Vercio, C., Loo, L. K., Green, M., Kim, D. I., & Beck-Dallaghan, G. L. (2021). Shifting focus from burnout and wellness toward individual and organizational resilience. *Teaching and Learning in Medicine, 33*(5), 568–576.
17. Wahab, E. (2025). Foresight study on the adoption of artificial intelligence (AI) in employee wellness program in Malaysia. *Research in Management of Technology and Business, 6*(1), 33–48.
18. Mantello, P., & Ho, M. T. (2024). Emotional AI and the future of wellbeing in the postpandemic workplace. *AI and Society, 39*(4), 1883–1889.
19. Tito, S. A., Arefin, S., et al. (2025). Integrating AI Chatbots and wearable technology for workplace mental health: Reducing stigma and preventing burnout through human-AI collaboration. *Central India Journal of Medical Research, 4*(01), 60–68.
20. Johnson, T., & Shamroukh, S. (2025). Assessing the predictive influence of organizational culture on employee burnout within health systems: Insights and strategic implications. *Artificial Intelligence in Health, 34*, 5127.
21. Advithi, M. J., & Ganesha, K. S. (2025). Measuring the impact of employee well-being on employee retention using a machine learning approach. In *Insights into digital business, human resource management, and competitiveness* (pp. 207–224). IGI Global Scientific Publishing.
22. Bokhari, M. U., Yadav, G., & Zeyauddin, M. (2024). Innovating workplace mental health strategies with advanced machine learning: Application of a superior ensemble classifier for accurate stress detection in business environments. *International Journal of Information Technology, 131*, 1–16.
23. Castel-Branco, G. D. L. V. (2024). Bridging technology and well-being: Applying machine learning to employee mental health: An investigation into the integration, impact, and adoption of intelligent solutions (Master's thesis, Universidade NOVA de Lisboa (Portugal)).
24. Singh, B. (2025). Assimilating employee well-being into sustainable business practices for successful organization: A path to long-term growth and productivity. In *Prioritizing employee mental health and well-being for organizational success* (pp. 161–184). IGI Global Scientific Publishing.
25. Dey, D. N. C. (2024). Harmonizing the threads of well-being: AI (artificial intelligence), spirituality, healthcare, and management. *Spirituality, Healthcare, and Management.*
26. Melnyk, B. M., & Amaya, M. (2024). Building and sustaining organizational wellness cultures to enhance clinician. *Evidence-Based Leadership, Innovation, and Entrepreneurship in Nursing and Healthcare: A Practical Guide for Success, 142*, 5565.
27. Zawaideh, F., & Bataineh, B. (2025). ICAIMT-machine learning-driven innovations in knowledge management: Integrating business intelligence, e-learning and organizational culture for AI management trends. *Journal of Information and Knowledge Management, 43*, 2550057.
28. Li, K., Mai, F., Shen, R., & Yan, X. (2021). Measuring corporate culture using machine learning. *The Review of Financial Studies, 34*(7), 3265–3315.
29. Ylinen, M., & Ranta, M. (2021). Employee-friendly corporate culture and firm performance: Evidence from a machine learning approach.
30. Gordon, G. G., & DiTomaso, N. (1992). Predicting corporate performance from organizational culture. *Journal of Management Studies, 29*(6), 783–798.
31. Miao, Y., & Zhang, W. (2024). Big data analysis and model construction of the correlation between organizational culture and employee performance. *Journal of Logistics Information Service Sciences, 11*(1), 416–428.
32. Hassan, F. S. U., Shah, B., Ikramullah, M., Zaman, T., & Khan, H. (2011). The role of organization culture in predicting organizational effectiveness: A case from developing countries. *International Business and Management, 3*(2), 99–111.
33. Jafari, M., Mehdipour, Y., Vandad, O., & Fallah, T. (2024). A prediction model for organizational performance based on leadership styles and the mediating role of innovative culture and organizational learning. *Journal of Psychological Science, 23*(135), 195–209.

34. Song, M., & Gao, L. (2024). Correlation analysis of corporate culture and employee behavior: A predictive model based on deep learning. In *International conference on big data technology and business analytics* (pp. 80–90). Springer.
35. Oredo, J., Njihia, J., & Iraki, X. N. (2017). The role of organizing vision in cloud computing adoption by organizations in Kenya.

Chapter 7
Ethical Considerations and Challenges

AI-driven wellness initiatives raise important ethical issues and obstacles, especially related to data privacy, algorithmic bias, and transparency. Although these programs present possible advantages, such as tailored insights and better healthcare access, they also raise issues regarding the responsible handling of sensitive information, the risk of biased suggestions, and the necessity for clarity in AI-driven decisions. The transformation and enhancement of long-term care systems and health results face notable obstacles within existing healthcare regulations. Recent technological progress has sparked interest in innovative healthcare applications, making this sector a burgeoning area for research. AI-based monitoring and decision support systems appear to have potential in expanding personalized care programs to remote and home environments [1]. The World Health Organization (WHO) acknowledges the essential function of technology in effectively incorporating long-term healthcare services into the daily routines of patients. Technology is deemed essential for attaining universal health coverage across all age groups by facilitating efficient integration and service delivery. Technological innovations, especially in AI and robotics, are currently utilized to aid healthcare research and practice. In the last ten years, the exploration of the use of AI to improve essential clinical processes and results has consistently increased. Decision support systems powered by AI can enhance clinical workflows, increase patient safety, assist in diagnosis, and facilitate personalized treatment. Numerous digital technologies, such as intelligent mobile and wearable devices, have been created to collect data and analyse information for evaluating health and monitoring personalized treatment advancements. Furthermore, social and physical support systems are being implemented to assist people in their rehabilitation from disease, injury, or sensory, motor, and cognitive disabilities. Data protection aims to safeguard personal information against unauthorized access by third parties, malicious threats, and misuse. It is designed to safeguard personal information through various methods and techniques to maintain data confidentiality [2].

S. Majumder and N. Dey, *AI-Driven Wellbeing to Enhance Lives and Work Environments*, Studies in Computational Intelligence 1257,
https://doi.org/10.1007/978-3-032-15735-5_7

7.1 Data Privacy and Security in Wellness AI

Data privacy and data security are interconnected yet different concepts. Data privacy emphasizes individuals' rights to manage their data and its collection, usage, and dissemination. Conversely, data security emphasizes safeguarding data from unauthorized access, breaches, or loss. In essence, data privacy specifies what information is gathered and its intended use, whereas data security describes the measures taken to safeguard that information [3]. Data security guarantees the integrity of information, ensuring that data are precise, dependable, and accessible to permitted individuals. Cybersecurity includes both computer and digital domains and involves a range of protective strategies aimed at securing our digital resources from possible dangers, such as human and technical mistakes, malicious individuals, and unauthorized access. This is precisely where the term information security becomes relevant, referring to the act of averting unauthorized access, use, disclosure, alteration, or destruction of information in any format.

Maintaining data privacy and security is essential in wellness initiatives to safeguard confidential employee health data. Organizations need to comply with regulations such as the HIPAA and GDPR, establish strong security protocols, and promote transparency to cultivate trust among employees. Wellness initiatives must comply with applicable data privacy regulations such as the Health Insurance Portability and Accountability Act (HIPAA) and the General Data Protection Regulation (GDPR). Restrict data gathering to only what is essential for the program and enforce stringent access controls to guarantee that only authorized individuals can access sensitive data. Encryption and secure storage methods are utilized to safeguard data against unauthorized access or breaches. Informed consent from employees should be secured before their data are gathered and utilized, and how their data will be utilized and safeguarded should be explicitly conveyed. When engaging third-party vendors, they perform comprehensive due diligence to confirm that they maintain robust security protocols, adhere to applicable regulations, and advise CoreHealth. A straightforward procedure for managing data breaches, which includes informing affected individuals and relevant authorities, has been created [4, 5]. An environment of openness and clear communication regarding data privacy and security measures should be cultivated to establish trust with employees.

Wellness initiatives frequently necessitate that employees share their personal information via biometric screenings and health risk evaluations (HRA). In certain situations, especially with group health plans, insurance claim information might also be accessible to HR teams, external wellness programs, and company leaders [6].

7.1.1 Understanding Data Privacy Regulations in Health Apps

Daily steps can be monitored via a health app, which can generate a sense of pride in achievements, until one learns through a news article that the app shares information with external parties. Such a disclosure can be startling, considering the extent of personal data voluntarily shared for the sake of convenience. With the growth of health apps, the importance of comprehending data privacy regulations must be acknowledged. Regulations such as the HIPAA in the US and the GDPR in Europe require companies to protect health data. Nevertheless, many users still have a limited understanding of their rights and the effects of these regulations on data security.

In the workplace, the involvement of employee wellness apps with these privacy regulations must also be taken into account. Like individuals require protection, organizations must ensure compliance to uphold trust. For example, tools such as Vorecol can offer insights into workplace morale while ensuring that employee privacy is protected. An effective method for assessing organizational climate is highlighted while complying with data regulations, allowing companies to cultivate a positive work environment without the concern of data breaches [7]. In a realm where data are regarded as money, understanding these protections is not just recommended but crucial.

7.1.2 Common Security Threats to Wellness Applications

At some points, personal health information may be retrieved from a wellness app that was believed to be safe. Such an event is not as improbable as it might appear. Recent studies indicate that security vulnerabilities may exist in almost 50% of wellness applications, jeopardizing private data. Hackers continuously look for vulnerabilities, and as digital solutions for health and wellness become more widely used, the potential targets for attacks grow significantly [8]. Importantly, although these apps can offer substantial advantages for well-being, they might also subject users to various security risks, such as data breaches, phishing scams, and vulnerable APIs.

This scenario emphasises the crucial importance of obtaining user consent in data management. Businesses must now focus more on gathering, handling, and safeguarding user data. As regulations change and consumer awareness increases, not obtaining informed consent could lead to financial penalties as well as considerable harm to brand trust. In the era of data-informed decision-making, organisations must foster a culture of transparency. Users must be provided with clear choices regarding data usage, ensuring that they remain informed and that their rights are respected [9]. A constructive workplace culture that highlights ethical data management practices can be promoted with the support of tools such as Vorecol's work environment. By facilitating the assessment of employee sentiment and involvement, a more cooperative environment can be fostered where consent and ethics are prioritised. In the end, when users are confident that their data is managed responsibly, successful businesses and enduring customer relationships can be established.

7.1.3 Best Practices for Securing Health Data

As one steps into a hospital, lines of computers holding confidential patient data can be observed. A year ago, a significant healthcare provider reported a data breach impacting more than 3 million patients [10]. A surprising question emerges: what measures can healthcare organizations take to safeguard such vital information? In the current digital era, optimal practices for protecting health data are more crucial than ever. Robust password policies, frequent software updates, and encryption methods are considered essential measures. Furthermore, a culture of awareness should be developed among staff to turn the workforce into the primary line of defense against possible threats.

Nevertheless, additional factors beyond just technical solutions need to be taken into account. The significance of the work environment in influencing security measures should be acknowledged. Compliance with protocols is more likely to be guaranteed when a supportive atmosphere is upheld. The mental health of employees can be assessed via instruments such as the Vorecol work environment. When team members experience value and engagement, security protocols intended to safeguard patient data are less likely to be ignored. Hence, by integrating strong technical approaches with a nurturing workplace environment, health data can be protected, and overall organizational confidence can be improved [11, 12].

7.1.4 The Impact of Data Breaches on Patient Trust

Upon entering a doctor's office, a reception area buzzing with activity may be observed. Patients are seen checking in, yet an underlying tension is felt. Recent news headlines about healthcare data breaches have been reported, and many individuals feel vulnerable. Studies have shown that nearly 70% of patients lose trust in their healthcare providers after a data breach. This statistic highlights the fragility of the relationship between patients and healthcare organizations, as trust can be eroded instantaneously. What is at stake is not only stolen information but also emotional fallout, through which the integrity of providers is questioned by patients. With trust shaken, reputations are required to be rebuilt by healthcare organizations with greater effort than ever before. One effective approach is the active measurement and improvement of the workplace environment in which staff and patient interactions occur. Tools such as Vorecol's work environment module are employed for this purpose. Through the measurement of employee satisfaction and the cultural climate, a more secure and reliable atmosphere can be fostered, extending to patient care. When support and engagement are felt by staff, their treatment of patients is positively reflected, fostering trust and ensuring that data security is prioritized within a rapidly evolving digital landscape [13, 14].

7.1.5 Emerging Technologies for Enhanced Privacy Protection

Upon logging into a preferred social media platform, a notification might appear indicating that the data are now confidential and exclusively available to the user. Even if it seems idealistic, the potential is realized through new technologies such as blockchain and zero-knowledge proofs. These advancements have led to the creation of decentralized systems that maintain the confidentiality of personal information while granting users increased control over access to their data. A recent study revealed that 83% of consumers express worries regarding online privacy, highlighting the pressing requirement for solutions to address these concerns. In professional settings, the significance of privacy-focused tools is growing, especially as remote work situations test conventional data protection methods. One can envision a platform that assesses employee satisfaction while ensuring that all feedback remains confidential in a protected cloud setting. Vorecol's work environment module is being implemented to provide companies with insights into employee morale while keeping personal information private [15]. Such technologies can promote trust and transparency, allowing organizations to succeed while maintaining employee privacy. In a world that places greater importance on data security, these advancements are welcomed as possible solutions for a safer future.

Equally troubling situations are observed in health technology. A scenario where a health application, intended to monitor heart rate and sleep patterns carefully, is breached, resulting in private personal information being accessed by cybercriminals. These risks are increasing, as health applications are expected to surpass a market value of $200 billion by 2026. As dependence on digital tools for health management grows, the need for strong cybersecurity measures is being acknowledged. Trends such as biometric authentication and end-to-end encryption are becoming standard practices to safeguard the privacy and security of personal health data.

With the ongoing evolution of regulations such as the GDPR and HIPAA, developers of health apps face increasing pressure to adhere to strict data protection requirements. Companies' focus on security not only protects user information but also enhances trust, which has become a crucial asset in the personal health technology sector. The importance of fostering a safe organizational atmosphere is highlighted by tools such as Vorecol for workplaces. By assessing employee satisfaction and engagement, a culture of security awareness is cultivated, enhancing compliance initiatives. With the merging of health technology and more stringent security measures, personal health management is likely to be transformed, guaranteeing safer and more dependable interactions with health applications.

The convergence of data privacy and security in health and wellness applications creates a complicated environment that requires immediate focus from developers, policymakers, and users [16, 17]. With these applications collecting more sensitive personal data, the risks linked to data breaches and unauthorized access are becoming increasingly serious. Users frequently have significant faith in these digital tools, as they believe that their health information is protected when, in fact, numerous apps are deficient in strong security protocols and clear privacy policies. Therefore, a joint endeavour is crucial to create more robust regulatory frameworks and industry standards that emphasise user privacy while allowing innovation to thrive. Additionally, the responsibility does not rest only with app developers and regulators; users need to develop an understanding of the privacy risks associated with health and wellness apps. By comprehending how their data are utilized, distributed, and safeguarded, people can make more knowledgeable decisions about the applications they use and the data they disclose. Promoting open dialogue regarding data practices and establishing user education programs are essential measures in cultivating a culture of privacy awareness within this growing industry. In the end, tackling the issues of data privacy and security will not only build user trust but also open doors for the ongoing development and success of health and wellness technologies in a responsible way.

7.2 Addressing Bias in AI Wellness Solutions

Tackling bias in AI wellness solutions necessitates a comprehensive strategy that involves varied and representative data, tools for detecting bias, human-in-the-loop systems, and continuous monitoring and assessment. By actively addressing bias, we can ensure that AI wellness solutions are just, equitable, and advantageous for every user. The basis of an impartial AI system is the information it learns from. It is essential to guarantee that training datasets incorporate a diverse array of demographics, traits, and healthcare requirements to truly represent the target population [18]. Using fairness metrics, explainable AI (XAI), and adversarial testing can assist in detecting and correcting biases in AI models. The incorporation of human experts, particularly clinicians and data scientists, can aid in detecting possible biases and guarantee that AI solutions focus on clinical usefulness rather than solely on technical performance.

Consistently assessing the performance of AI systems and measuring them against predefined criteria (accuracy, sensitivity, specificity, etc.) is crucial for ensuring fairness and identifying new biases. External experts conducting independent audits of AI systems can assist in pinpointing and rectifying biases, promoting fairness and precision in various healthcare environments. Informing both healthcare providers and patients about possible biases in AI is essential for cultivating a mutual understanding and encouraging responsible AI use in healthcare [19]. Cognitive biases: These are unintentional mistakes in reasoning that influence people's evaluations and choices. These biases stem from the brain's effort to streamline the processing of information regarding the world. Psychologists have identified and categorized over 180 human biases. Cognitive biases may infiltrate machine learning algorithms through designers inadvertently embedding them into the model or a training dataset that contains those biases. Algorithm bias: AI technologies or machine learning software perpetuate existing biases found in the training data or through the design of the algorithm. This may occur because of explicit biases in the coding or existing beliefs held by the creators. For example, a model that places too much importance on income or education can perpetuate negative stereotypes and discrimination toward marginalized communities. Incomplete data: When data are lacking, they may not accurately represent the population and could therefore introduce bias [19].

Figure 7.1 shows the biases inherent in the machine learning (ML) pipeline, highlighting how different phases of an ML workflow can introduce multiple forms of bias. The procedure is split into five phases, each linked to typical difficulties. During the initial phase of defining the research problem, various biases may surface, including racial or ethnic bias, gender bias, age bias, disability bias, English as a second language (ESL) bias, and global misrepresentation bias, as the scope and goals are established. During data collection, challenges such as sampling bias, measurement bias, exclusion bias, label bias, and social bias may occur due to the methods used for data acquisition. During the data preprocessing phase, biases such as aggregation bias, feature selection bias, outlier bias, and confounding bias can occur while organizing or preparing the data. In the process of model development and validation, biases may include training dataset bias, test dataset bias, algorithmic bias, confirmation bias, and validation bias, indicating constraints in both the data and methodologies. Ultimately, when the model is implemented, new difficulties such as concept drift and covariate shift can arise as real-world circumstances evolve. This illustration highlights that bias can emerge at any phase of the ML pipeline—from defining the problem to deployment—and needs to be approached proactively. Effectively addressing these biases is crucial for developing AI systems that are just, precise, and ethical.

7.3 Ensuring Accessibility and Inclusivity

Artificial intelligence (AI) can enable people from various backgrounds, skill levels, and communities. To realize its full potential, AI needs to be created and implemented with accessibility and inclusivity as fundamental priorities. This guarantees that every user, irrespective of physical capability, socioeconomic level, gender, culture, or location, can gain fairly from AI-enabled technologies. Accessibility emphasizes the creation of AI systems that can be utilized by individuals with disabilities or constraints. AI-based tools such as voice assistants, screen readers, speech-to-text, and image recognition assist users who are visually impaired, hearing impaired, or have mobility challenges [20, 21]. AI technologies can customize interfaces according to individual requirements (e.g., enlarged fonts, streamlined navigation, or different input methods). NLP supports translation, transcription, and multilingual

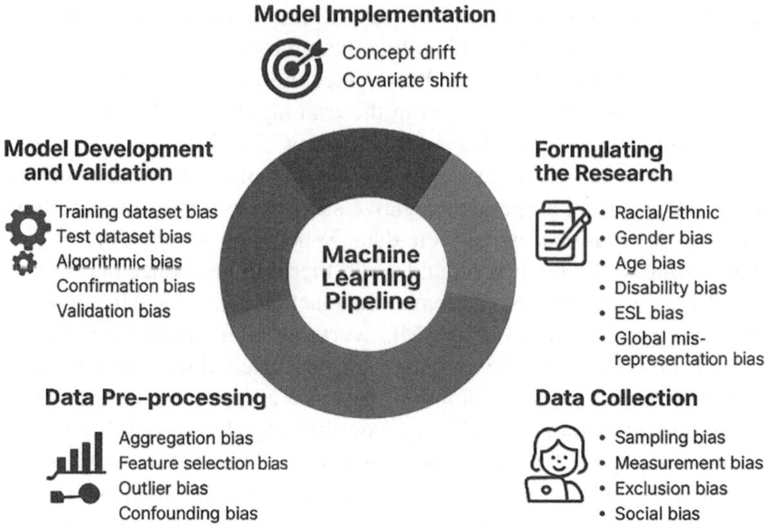

Fig. 7.1 Machine learning (ML) pipeline

capabilities, facilitating communication across languages. Accessibility and inclusivity in AI are vital; they are crucial for creating a fair and equitable digital community. Integrating fairness, diversity, and usability into AI systems allows people to develop technologies that support all individuals and communities, narrowing gaps rather than expanding them.

Figure 7.2 shows a Venn diagram illustrating the relationships among inclusivity, accessibility, and the POUR principles (perceivable, operable, understandable, robust). Together, they emphasize how digital systems and AI can be crafted for justice, user-friendliness, and equality. The essential components of inclusivity include acknowledgement, parity, and involvement. Concentrates on ensuring that every person, irrespective of their background, identity, or ability, feels acknowledged and appreciated. It is important to promote equal involvement and active participation in digital or AI-based settings. The essential components of accessibility are usability, adaptability, and comprehensibility. Guarantees that systems and technologies are accessible to individuals with various abilities. Accessibility eliminates obstacles by creating customizable tools (e.g., text-to-speech, captions, screen readers) that are comprehensible to every user. The essential components of the POUR Principles are perceivable, operable, understandable, and robust. The fundamental principles of web accessibility guidelines (WCAAs) guarantee that users can perceive information, interfaces can be operated with various input methods, content is comprehensible for a wide range of audiences, and systems are resilient, functioning across different devices and assistive technologies. The figure highlights that accessibility guarantees usability and that inclusivity guarantees fairness and participation, whereas POUR principles offer technical standards. Convergence ensures

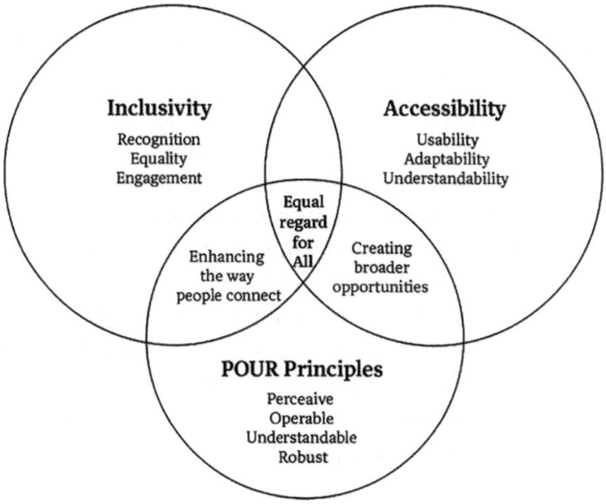

Fig. 7.2 Inclusivity versus accessibility

equal consideration for everyone, rendering technology universally accessible and advantageous.

7.3.1 Accessibility and Inclusivity: Ethical Considerations and Challenges

Guaranteeing accessibility and inclusivity in technology is an ethical obligation and a practical need in the era of AI-powered systems [22]. Ethical issues emerge when digital solutions neglect individuals with diverse abilities, backgrounds, and languages, thereby perpetuating inequality and digital exclusion. For instance, disregarding accessibility guidelines might marginalize people with disabilities, whereas cultural or language biases in AI systems can exclude nondominant communities. Inclusivity transcends mere compliance; it requires the proactive creation of systems that acknowledge and honour diverse needs, ensuring fair participation in education, work, healthcare, and public services [23]. The difficulty is in reconciling technical practicality with ethical obligations, urging organizations to emphasize universal design, inclusive data representation, and ongoing stakeholder involvement. Incorporating accessibility and inclusivity into the foundation of ethical frameworks allows technology to approach its potential of empowering everyone instead of favouring just a select group. A significant obstacle exists in how data are represented [24]. When training datasets inadequately represent individuals with disabilities, marginalized languages, or various sociocultural backgrounds, AI models might continue to reflect biases and strengthen existing systemic inequalities. For example, voice recognition software frequently has difficulty with accents or speech disabilities, and hiring algorithms might unintentionally exclude applicants with unconventional career trajectories or gaps due to caregiving or disabilities.

Another ethical concern is universal design versus custom design. While universal design principles aim to ensure that systems are accessible to the widest possible audience, they may not sufficiently address the specific needs of individual groups. Reaching the right balance requires engaging all stakeholders, including affected communities, in the creation and assessment of technologies. Ultimately, inclusivity also calls for the consideration of affordability and digital literacy. Although systems may be technically accessible, financial or educational obstacles can hinder fair participation [25]. Consequently, developing ethical technology must incorporate policies, design criteria, and ongoing oversight that foster accessibility, inclusivity, and equity. The incorporation of accessibility and inclusivity at the heart of technological advancement allows organizations to surpass basic compliance and adopt a human-centered strategy that promotes empowerment, equity, and social justice.

7.3.2 Future Trends in Workplace Wellness

In the future, workplace wellness, accessibility and inclusivity will be seen not as optional extras but as essential foundations for organizational success. With workforces becoming increasingly diverse and linked globally, organizations need to create wellness programs that cater to different physical, cognitive, cultural, and linguistic requirements [26, 27]. A significant trend will be the growth of AI-driven personalization, where wellness platforms employ adaptive algorithms to customize resources, suggestions, and interventions for employees with varying abilities, backgrounds, and health statuses.

Accessibility enhanced by technology will grow through assistive tools such as voice-activated platforms, AI translators for diverse workforces, and wearables that monitor individualized health information while supporting disabilities. Additionally, organizations adopt universal design principles, guaranteeing that digital wellness tools are created to benefit all users from the beginning instead of making adjustments afterward. A rising trend is the incorporation of neurodiversity-friendly practices, with wellness programs assisting individuals with ADHD, autism, or other cognitive differences through adaptable scheduling, mindfulness training, and tailored learning resources [28, 29]. Wellness inclusivity encompasses mental health and emotional well-being, dismantles stigmas, and addresses obstacles encountered by marginalized groups such as LGBTQ + employees, remote workers in secluded settings, and individuals managing caregiving duties [30]. Moreover, companies will progressively aim for fair access, guaranteeing that wellness benefits reach not only affluent employees or those in corporate settings but also gig workers, part-time workers, and staff in developing markets [31]. Ultimately, accessibility and inclusivity transform workplace wellness into a comprehensive, fair, and human-centered environment where technology, policy, and culture work together to guarantee that all employees, irrespective of ability, identity, or background, can flourish.

7.4 Conclusions

The swift progress of artificial intelligence and associated technologies has generated unmatched opportunities while also revealing significant ethical issues. This chapter emphasized that concerns such as bias, transparency, accountability, privacy, and fairness are not just technical problems but also fundamental societal issues. The ethical aspects of AI require a thoughtful equilibrium between progress and accountability, guaranteeing that advancements do not jeopardize human dignity, inclusiveness, or fairness. The challenges involve not only recognizing and reducing risks but also creating strong governance structures, fostering organizational consciousness, and integrating ethics into the design and implementation of AI systems. Crucially, ethics should be viewed as an ongoing process instead of a one-off compliance task, accommodating new contexts, information, and societal demands.

References

1. Thompson, A., Stringfellow, L., Maclean, M., & Nazzal, A. (2021). Ethical considerations and challenges for using digital ethnography to research vulnerable populations. *Journal of Business Research, 124*, 676–683.
2. Eden, C. A., Chisom, O. N., & Adeniyi, I. S. (2024). Integrating AI in education: Opportunities, challenges, and ethical considerations. *Magna Scientia Advanced Research and Reviews, 10*(2), 006–013.
3. Safdar, N. M., Banja, J. D., & Meltzer, C. C. (2020). Ethical considerations in artificial intelligence. *European Journal of Radiology, 122*, 108768.
4. Das, S., Chakraborty, S., Sajjan, G., Majumder, S., Dey, N., & Tavares, J. M. R. (2022). Explainable AI for predictive analytics on employee attrition. In *International conference on soft computing and its engineering applications* (pp. 147–157). Springer.
5. Gupta, A., Raj, A., Puri, M., & Gangrade, J. (2024). Ethical considerations in the deployment of AI. *Tuijin Jishu/Journal of Propulsion Technology, 45*(2), 1001–4055.
6. Visave, J. (2024). AI in emergency management: Ethical considerations and challenges. *Journal of Emergency Management and Disaster Communications, 5*(01), 165–183.
7. Baeza-Yates, R. (2022). Ethical challenges in AI. In *Proceedings of the fifteenth ACM international conference on web search and data mining* (pp. 1–2).
8. Prajapati, S. B. (2025). Ethical considerations in AI design and deployment. *World Journal of Advanced Research and Reviews, 25*(1), 2166–2173.
9. Kumar, A. (2024). The role of synthetic data in advancing AI models: Opportunities, challenges, and ethical considerations. *Journal of Artificial Intelligence General Science (JAIGS), 5*(1), 443–459.
10. Hua, S., Jin, S., & Jiang, S. (2024). The limitations and ethical considerations of Chatgpt. *Data Intelligence, 6*(1), 201–239.
11. Vermesan, O., De Luca, C., John, R., Coppola, M., Debaillie, B., & Urlini, G. (2022). Ethical considerations and trustworthy industrial AI systems. *Intelligent Edge-Embedded Technologies for Digitizing Industry, 31*, 1542.
12. Patel, D., Chetarajupalli, C., Khan, S., Khan, S., Patel, T., Joshua, S., & Millis, R. M. (2025). A narrative review on ethical considerations and challenges in AI-driven cardiology. *Annals of Medicine and Surgery, 87*(7), 4152–4164.
13. Singh, B. P., & Joshi, A. (2024). Ethical considerations in AI development. In *The ethical frontier of AI and data analysis* (pp. 156–179). IGI Global Scientific Publishing.

14. Ganesan, A. (2025). Transforming wellness: Ethical, cognitive, and privacy challenges in AI-driven healthcare. In *Harnessing AI and machine learning for precision wellness* (pp. 481–500). IGI Global Scientific Publishing.
15. Kostopoulos, L. (2018). The emerging artificial intelligence wellness landscape: Benefits and potential areas of ethical concern. *Calculation WL Review, 55*, 235.
16. Oluwagbade, E. (2025). Ethical AI in employee wellness: Balancing data-driven insights with employee privacy and trust.
17. De Freitas, J., & Cohen, I. G. (2024). The health risks of generative AI-based wellness apps. *Nature Medicine, 30*(5), 1269–1275.
18. Ang, C. S., & Lee, K. F. (2025). Ethical considerations in AI-powered health communities to promote good health and wellbeing: A narrative review. *Navigating Health Information in the Age of Artificial Intelligence, 32*, 67–94.
19. Rahsepar Meadi, M., Sillekens, T., Metselaar, S., van Balkom, A., Bernstein, J., & Batelaan, N. (2025). Exploring the ethical challenges of conversational AI in mental health care: Scoping review. *JMIR Mental Health, 12*, e60432.
20. Dey, D. N. C. (2024). Harmonizing the threads of well-being: AI (Artificial Intelligence), spirituality, healthcare, and management. *Spirituality, Healthcare, and Management, 64*, 1134.
21. Şendur, E. G., & Turgut, A. (2020). Artificial intelligence in older adults' healthcare: Applications and ethical considerations. *Akdeniz Hemşirelik Dergisi, 4*(2), 90–99.
22. Thakur, R. A., Talukdar, M., Iyer, S., Liu, L. C., Chen, C. L., & Singh, A. (2025). AI and employee well-being: Assessing the ethical implications of AI-driven human resource practices in Indian universities. *Journal of Ecohumanism, 4*(1), 2338–2351.
23. Naithani, K., Raiwani, Y. P., Tiwari, S., & Chauhan, A. S. (2025). Ethical aspects of AI nutrition: Discussion of ethical considerations related to AI-based nutritional guidance in terms of privacy, data security, and potential biases. In *AI-based nutritional intervention in polycystic ovary syndrome (PCOS)* (pp. 271–290). Springer.
24. Khaitan, J., Raza, A. M., & Ghosh, J. (2025). Introduction to precision wellness. In *Harnessing AI and machine learning for precision wellness* (pp. 1–40). IGI Global Scientific Publishing.
25. Majumder, S., & Dey, N. (2022). *AI-empowered knowledge management*. Springer.
26. Marko, J. G. O., Neagu, C. D., & Anand, P. B. (2025). Examining inclusivity: The use of AI and diverse populations in health and social care: A systematic review. *BMC Medical Informatics and Decision Making, 25*(1), 57.
27. Zallio, M., Ike, C. B., & Chivăran, C. (2025). Designing artificial intelligence: Exploring inclusion, diversity, equity, accessibility, and safety in human-centric emerging technologies. *AI, 6*(7), 143.
28. Ang, C. S., & Lee, K. F. (2025). Ethical considerations in AI-powered health communities to promote good health and wellbeing: A narrative review. *Navigating Health Information in the Age of Artificial Intelligence, 48*, 67–94.
29. Putrevu, D. J., & Mertzanis, C. (2025). The rise of the wellness business sector: A systematic literature review.
30. Alam, M. M., Mustafa, J., Tasnim, N., Siddik, A. B., & Islam, A. R. (2018) Technology-enabled employee wellness and sustainability: A literature-based review.
31. Ghosh, S., Majumder, S., & Peng, S. L. (2023). An empirical study on adoption of artificial intelligence in human resource management. In *Artificial intelligence techniques in human resource management* (pp. 29–85). Apple Academic Press.

Chapter 8
Future Trends in AI for Holistic Wellness

Artificial intelligence (AI) is swiftly changing every aspect of our existence, and the area of holistic wellness is no different. As individuals progress into the twenty-first century, the integration of AI and wellness is set to transform how people consider physical, mental, emotional, and spiritual well-being. Holistic wellness, which highlights the link between the mind, body, and environment, greatly benefits from AI's ability to analyse extensive datasets, tailor experiences, and forecast results with impressive accuracy [1]. New trends in AI are facilitating creative solutions such as individualized health management, mental wellness assistance, dietary advice, and mindfulness training, all customized to the specific requirements of every person. Additionally, AI-powered platforms promote a more cohesive approach to wellness, linking users with professionals, resources, and communities instantly. This investigation into upcoming trends in AI for comprehensive wellness will examine the advanced technologies and methods influencing the next era of well-being—emphasizing how AI can enable individuals to attain enhanced balance, resilience, and satisfaction in their everyday lives. Upcoming trends in AI for comprehensive wellness will focus on individualized, proactive, and preventive care, fuelled by developments in AI-driven predictive analytics, virtual health aides, and AI-enhanced wearable devices. These technologies improve physical and mental health by providing personalized suggestions, early illness identification, and easy access to assistance [2]. AI algorithms examine large volumes of data from wearables, health records, and lifestyle practices to forecast possible health risks and provide tailored suggestions for nutrition, physical activity, and stress reduction. This proactive strategy seeks to avert diseases prior to their appearance, enhancing their general health. AI-driven ECG monitoring in wearable devices can identify atrial fibrillation at an early stage, enabling prompt intervention. Virtual health assistants, such as chatbots and AI-driven coaches, offer immediate health guidance, encouragement, and assistance in maintaining a healthy lifestyle. They can arrange appointments, remind people to take their meds, and provide tailored advice on fitness, nutrition, and mental wellness. This enhanced availability and ease can diminish the stigma linked to

S. Majumder and N. Dey, *AI-Driven Wellbeing to Enhance Lives and Work Environments*, Studies in Computational Intelligence 1257,
https://doi.org/10.1007/978-3-032-15735-5_8

asking for assistance and foster greater participation in wellness activities [3]. AI has been progressively utilized to increase mental wellness via tailored interventions, early identification of mental health problems, and therapy access through online platforms. Tools driven by AI can examine emotional data from multiple sources to detect patterns and offer personalized assistance for individuals facing anxiety, depression, or other mental health issues. VR and AR technologies are utilized to develop immersive and soothing settings for relaxation and meditation [4].

8.1 Advancements in AI-Driven Wellness Technologies

The wellness sector has experienced a significant transformation due to the emergence of artificial intelligence (AI). Because they were previously dependent on standard wellness programs and uniform methods, contemporary wellness efforts now utilize AI to provide customized solutions aligned with a person's distinct requirements. This technological progress has made health and wellness more reachable, efficient, and impactful than previously. AI-powered smartwatches and fitness trackers have revolutionized the way people track their health. These devices monitor steps, heart rate, sleep quality, and even stress levels. AI algorithms examine these data to deliver customized feedback and practical suggestions [5]. Practical instances include Apple Watch and Fitbit, which employ AI to identify abnormal heart rhythms and forecast possible health issues, such as atrial fibrillation, and encourage users to consult a doctor before symptoms worsen. AI-powered wellness platforms such as Noom and Lumen utilize machine learning to provide tailored coaching. These platforms monitor dietary behaviors, workout routines, and metabolic rates to offer personalized health recommendations. By 2025, innovations powered by AI are reshaping how we address our health, fitness, and emotional wellness. Reports from tech media emphasize the surge of AI tools designed for mental health assistance and tailored healthcare, indicating that what was previously considered futuristic is now becoming commonplace. Simultaneously, innovative studies have demonstrated progress in AI algorithms capable of identifying diseases sooner, tailoring treatments, and enhancing the efforts of human caregivers.

I. **AI-Powered Mental Health Support**

AI is progressively increasing mental health support with chatbots and virtual aids. Educational institutions and workplaces have started using hybrid, human–AI wellness chatbots to assist individuals who do not have access to counselors. These AI services offer round-the-clock text support, allowing users to express their worries or rehearse challenging discussions in a compassionate, noncritical environment. Services such as Clare&me in Germany and Limbic Care in the UK provide round-the-clock AI "therapy companions" for individuals who may not have access to assistance [6, 7]. These AI tools focus on social-emotional aspects, interact with users, track their mental health, and guide them to appropriate resources. Preliminary studies indicate that AI-based chatbots may offer compassionate, stigma-free assistance for conditions such as anxiety and depression, although their efficacy is still being assessed. Nevertheless, specialists warn that ethical boundaries are essential. Concerns such as privacy, data bias, and the drawbacks of "chatbot therapy" need to be addressed to safely incorporate AI into mental health care. A collaborative research effort between OpenAI and the MIT Media Lab revealed a link between greater daily use of AI chatbots and heightened feelings of loneliness and reliance, indicating that AI chatbots could produce unintentional consequences. By 2025, the trend is leaning towards AI being used as scalable support for mental health and as a tool for clinical decision-making, broadening access to care while enhancing the efforts of human therapists [8, 9].

II. **AI in Healthcare and Medicine**

By 2025, advancements in healthcare research, diagnostics, and treatment will be propelled by AI. Generative AI, specifically, is revolutionizing the process of drug discovery. A leading researcher at Nvidia and Caltech indicated that the research and consulting company Gartner forecasts that more than 30% of new medications will be identified via generative AI by 2025 [10]. Scientists are currently utilizing AI systems to create new drugs in months instead of years. At Stanford University, researchers created a generative model called SyntheMol, which generated chemical "formulas" for six potential antibiotic candidates aimed at fighting a dangerous superbug, with findings published in Nature.

Outside the laboratory, AI is enhancing clinical decision-making. Sophisticated algorithms are capable of examining medical images and patient information with impressive precision. A Lancet Digital Health study from 2025 indicated that mammography screening supported by AI identified 29% more breast cancers, which includes 24% more early-stage tumors, than standard screening methods. AI is being utilized in neurology and preventive care: a study revealed that speech-analysis AI could predict Alzheimer's disease with almost 80% accuracy six years before diagnosis. Moreover, "ambient" AI systems in hospitals currently record visits and prepare clinical notes to alleviate physician burnout. Medical professionals and healthcare technology leaders emphasize that these tools need to be thoroughly validated and utilized responsibly [9]. Regulators are starting to guide AI "copilots" in healthcare to guarantee safety and transparency. In summary, 2025 signifies a turning point as AI evolves from experimental concepts to real-world healthcare uses, accelerating drug development, improving early disease identification, and assisting clinicians in providing more accurate and preventive care [11].

III. **Personalized Fitness and Wellness Tech**

AI has emerged as a fundamental element of contemporary fitness coaching, assessing performance metrics to modify workout plans in real time, thus providing an exceptionally tailored experience. The intelligent fitness devices revealed at CES 2025—ranging from smart mirrors to innovative wearables—were all equipped with AI-powered functionalities. For example, the newest smart rings offer "improved AI-driven intelligence" to monitor activity and heart health. Modern smart fitness devices, such as the Amp home gym and Gym Monster 2, employ AI to track form and offer tailored recommendations, simulating the support of a personal trainer [12]. A different device, the smart mirror, employs built-in sensors and AI to evaluate body composition, heart and lung health, and even dietary patterns every morning. While remaining a prototype, it suggests a future in which a bathroom mirror might serve as a personal health scanner and advisor [13].

AI is driving a new wave of personal training applications. Applications such as Juggernaut AI act as portable virtual coaches, designing personalized exercise routines that evolve with advancement. These applications gather performance metrics and objectives to create a custom workout plan, fine-tune sets and perform repetitions with AI-enhanced accuracy. A notable innovation is Whoop Coach, an AI-powered performance assistant that delivers real-time, conversational guidance customized to individual biometric information. In contrast to conventional wearables that merely gather data, Whoop Coach interprets intricate physiological signals—such as HRV, sleep, strain, and recovery—into clear, actionable recommendations. Pose inquiries such as "Is today a good day for training?" or "What caused my poor sleep?" and provide personalized advice, utilizing OpenAI's extensive language models alongside Whoop's exclusive dataset. This signifies a transition to adaptive wellness coaching, where the AI not only nudges or monitors but also transforms into a digital wellness guide that adapts to lifestyle. As wearables grow more personal (rings, patches, and even earbuds), more fitness brands are expected to incorporate LLM-driven AI companions, such as Whoop Coach, into their services [14, 15].

IV. **AI in Social Well-Being**

In addition to physical well-being, AI addresses the social needs of individuals. A notable trend in 2025 is the emergence of AI companions, digital or robotic entities that offer conversation, caregiving support, and companionship. Research shows that many older adults are receptive to human-like assistant robots, especially if this leads to increased independence and reduced feelings of loneliness. Contemporary companion robots, such as the ElliQ device, can answer voice commands, participate in casual conversations, remind users to take medications, recommend exercises or brain games, and even make calls to family and friends. Importantly, designers incorporate empathetic elements: these AI companions are designed to replicate human warmth and empathy, enhancing their attractiveness as social companions. For young individuals, the role of AI in education is expected to expand significantly, especially in delivering personalized learning experiences tailored to each student's needs. The algorithms of machine learning help pinpoint knowledge gaps, recommend educational materials, and promote an inclusive strategy for various learning styles. AI companion services such as Replika (a well-known AI friend application) and Xiaoice boast millions of users and are increasingly regarded as valid sources of emotional support rather than just novelties [16]. In a survey involving 1000 college students, Replika not only benefitted the majority of users but also helped prevent suicide attempts among 30 individuals. Through offering noncritical listening and instant messaging, AI companions are meeting social requirements for interaction and dialogue in a fast-paced environment—and potentially saving lives.

V. AI for Workplace Wellbeing

Organizations are progressively utilizing AI to foster healthier and more efficient workplaces. In the workplace, AI is expected to greatly improve productivity by automating mundane tasks, which permits employees to concentrate on more complex cognitive abilities. With the adoption of AI-powered productivity tools, research has shown that employee satisfaction and performance improve [17]. The possible advantages also correspond with a human-centered approach, emphasizing the importance of nurturing innovation and collaboration, thereby making human welfare a vital consideration in the application of these technologies. Moreover, certain organizations have started utilizing AI to oversee and enhance employee wellness. Companies such as Cisco and Adidas test AI tools that analyse employee feelings from messages, emails, and surveys, essentially performing an ongoing "vibe check" on morale and stress within the organization. By identifying early signs of decreased engagement or burnout, these systems can notify managers to intervene or modify workloads. The aim is to illuminate the frequently unseen elements of employee well-being (such as stress or loneliness, particularly in remote teams) to ensure that assistance can be offered quickly. Platforms for employee experiences powered by AI also collect feedback to recommend enhancements in the workplace, including policy changes and customized wellness resources designed for employees [18]. While just a small fraction of companies are presently utilizing AI for the "human aspect" of business, interest is increasing, as businesses understand that more content and healthier employees contribute to higher productivity and lower turnover rates.

Health plans powered by AI are notable for their ability to be tailored. In contrast to uniform workout programs or generic nutrition recommendations, these plans are developed via machine learning algorithms that analyse extensive user data, such as daily activities and exercise choices, dietary patterns, and health indicators. This analysis develops a customized plan for reaching health and fitness objectives more effectively. Machine learning allows AI to evaluate user data on a personal level, facilitating the development of workouts and nutrition plans that genuinely suit each individual's lifestyle. Through the examination of sleep patterns, activity levels, and recovery durations, machine learning algorithms can modify workout intensity or frequency to fit a person's requirements, fostering consistent advancement towards health objectives [19]. A significant role of AI in wellness is its capacity to foresee possible health threats. For example, if an individual's data show extended durations of increased heart rate and inadequate sleep, AI might highlight these measures as potential risk factors. This proactive strategy enables individuals to implement lifestyle changes before significant problems arise, ensuring that wellness encompasses not only fitness but also preventive health care.

AI has transformed the possibilities in wellness, allowing people to reach fitness and health objectives more efficiently and accurately. By analysing data, AI can determine the best methods to achieve a user's objectives, optimizing the route to success. Through AI-powered assistance, users steer clear of experimentation, attaining outcomes more quickly. In contrast to conventional coaching, AI provides ongoing assistance, keeping users motivated with immediate insights that guide every phase of their fitness journey [20]. AI analyses extensive data to uncover trends in user performance, offering insights crucial for enduring progress and aiding users in recognizing what suits them best. Customized plans foster dedication by regularly highlighting users' progress, helping them remain motivated to be active. AI has enhanced wellness and cessibility by providing affordable, scalable options that adjust as users' aspirations change, offering premium wellness support to a wider audience.

8.2 Integration with Virtual and Augmented Reality

Combining virtual reality (VR) and augmented reality (AR) technologies presents a bright outlook for engaging and interactive experiences, especially in mobile application development. By integrating the advantages of both technologies, developers can craft applications that harmoniously merge the physical and digital realms, boosting user interaction and offering creative solutions to real-life challenges. Augmented reality and virtual reality (AR & VR) currently represent some of the groundbreaking technological developments, and their ability to enhance the educational system is vast. The incorporation of augmented reality (AR) and virtual reality (VR) in education has increased significantly in recent years, offering numerous opportunities for technology-driven learning. AR and VR immerse students in digital experiences that traditional teaching methods cannot replicate, allowing for deeper engagement with complex subjects beyond mere lectures and textbooks while also allowing instructors to tailor content to accommodate various learning styles. These technologies not only enhance immersion but also enable educators to conduct simulations and undertake virtual field trips without the need for physical travel.

Augmented reality (AR) is a technology that improves the real-world environment by superimposing computer-generated elements onto it in professional settings across diverse sectors, such as healthcare, manufacturing, education, and retail. Organizations are now utilizing AR to increase employee safety through virtual training simulations and by visualizing the operation of equipment prior to its production [21]. Conversely, virtual reality (VR) is a cutting-edge technological advancement that has transformed our engagement and interaction with digital settings. By using advanced computer graphics, motion detectors, and display technologies, VR allows users to fully engage in strikingly realistic representations of actual or fictional environments. VR has demonstrated immense value across numerous sectors, including gaming, education, healthcare, and real estate. Although AR and VR may appear alike, they are two separate technologies that serve different functions. VR immerses individuals in an entirely different digital space, offering an engaging experience via headsets or goggles. AR enhances the real-world environment by superimposing digital objects onto it, adding additional information or improving its utility.

Additionally, AR and VR technologies provide various advantages for online learning, mobile learning, and blended learning, including immersive learning experiences and a more captivating learning atmosphere. AR and VR enable the creation of virtual environments and simulations that allow students to investigate and engage with actual surroundings without stepping outside the classroom. Moreover, AR and VR can be utilized to produce interactive and captivating content, such as 3D visuals and videos, which can aid in maintaining student interest. Additionally, the incorporation of AR and VR in educational settings has demonstrated a beneficial impact on student involvement and understanding. Students utilizing AR and VR technologies exhibit increased motivation and engagement, along with improved performance on academic assignments. This is probably due to AR and VR offering a more engaging learning experience, enabling students to investigate, uncover, and engage with their surroundings [22]. The implementation of AR and VR in educational settings has recently become increasingly popular, as schools and universities have started to invest in this technology. The price of AR and VR equipment has been gradually declining, making it accessible for individuals with limited finances. AR and VR technologies have also found applications in various educational environments, including museums, libraries, and science facilities. Through AR and VR, learners can investigate and engage with items in a manner that is impossible in the real world. This form of learning can be especially helpful for younger learners, who might find it difficult to understand intricate ideas without support from visual tools.

8.2.1 VR for Wellness

In the current rapid environment, the significance of mental health and wellness cannot be minimized. The workplace, which used to be centered primarily on productivity and efficiency, has transformed to acknowledge that the mental health of employees is fundamental to success.

Employers recognize that a team characterized by stress, anxiety, and societal pressures cannot operate at optimal levels. Consequently, they are more frequently seeking creative approaches to enhance employee mental health. VirtualSpeech involves the convergence of advanced technology and deep aspiration to enhance the mental well-being of both individuals and organizations [23]. While embarking on this journey, the aim is to shift the focus from conventional methods of workplace wellness, such as relaxation and visual escape, to a wider viewpoint that includes the transformative capabilities of virtual reality (VR) in tackling mental health challenges. In today's work environment, a distinct array of challenges for mental well-being arises. The demands of meeting deadlines, reaching goals, and sustaining a work–life balance frequently lead to stress, anxiety, and exhaustion.

I. **Stress Reduction**

VR can immerse employees in soothing virtual settings, such as natural scenes or tranquil landscapes, for guided meditation and relaxation practices, assisting in reducing workplace stress. Virtual reality (VR) serves as a powerful resource for alleviating stress by enveloping users in serene, structured settings that foster tranquility and awareness. VR therapy may decrease cortisol levels, lower heart rate, and enhance emotional resilience. Immersive VR experiences, such as tranquil coastlines or vibrant woodlands, offer a sensory getaway from everyday stress. VR meditation applications assist users in breathwork, visualization methods, and mindfulness activities, improving concentration and fostering relaxation. VR can establish secure, regulated settings that assist individuals in progressively facing and coping with phobias or traumatic experiences. VR enables individuals to exercise stress-relief methods in various virtual settings, assisting them in cultivating and enhancing their coping strategies. VR settings reduce distractions, enabling a more concentrated and efficient relaxation experience [24]. VR has the potential to gamify meditation, enhancing its engagement and accessibility, particularly for individuals who struggle with conventional meditation. Being in virtual natural settings (such as forests and beaches) can reduce stress and enhance mood. VR can be utilized to inform people about stress and anxiety, offering insights and enhancing self-awareness. VR provides employees with a method to distance themselves from stressful work situations and rehearse relaxation strategies, enhancing their overall well-being. Research indicates that VR can reduce the levels of cortisol (a stress hormone). VR can help people improve their coping strategies and handle stress more efficiently.

II. **Fitness and Physical Activity**

VR fitness applications deliver engaging workout experiences, offering tailored training plans and instantaneous feedback on technique and performance. Virtual reality (VR) transforms fitness by providing immersive and tailored exercise experiences. By creating immersive settings, VR can enhance the enjoyment of exercise and inspire users to remain active, resulting in better fitness results and overall wellness. Research indicates that VR can make workouts seem less strenuous, which may increase motivation and commitment to exercise programs. Numerous VR fitness games, such as Beat Saber, deliver vigorous cardio workouts, enhancing heart health and stamina. VR can assist users in performing strength training and flexibility workouts, enhancing overall physical fitness. VR has demonstrated potential in alleviating pain throughout rehabilitation and exercise, assisting individuals in maintaining their activity levels. Exercising in virtual reality may increase cognitive ability, possibly improving concentration, focus, and mood. Immersive virtual reality settings can aid in lowering stress and anxiety and enhancing mental health. VR can simulate job tasks and highlight possible ergonomic problems, enabling modifications to workstations and tools to avert injuries.

III. **Immersive Environments**

Virtual reality (VR) greatly improves immersive settings by creating a completely simulated digital realm where users can engage and experience a feeling of presence, fostering a more profound sense of involvement and connection [25]. This is accomplished via specialized headsets and tracking systems that provide a 360°, three-dimensional experience. VR immerses users in a simulated space, creating a sense of physical presence in the digital realm. This sense of presence is essential for immersive experiences, enabling users to engage with the virtual environment as if it were genuine. VR allows users to engage with the virtual environment intuitively, handle objects, navigate their surroundings, and even interact with other users (in multiuser settings). Immersive VR settings can serve educational needs, enabling students to rehearse intricate activities, such as surgeries or flight simulations, within a secure and regulated environment. This practical experience can result in enhanced learning results and heightened involvement. VR provides a novel aspect of entertainment, enabling users to immerse themselves in games, navigate virtual environments, and engage in experiences they may not encounter in reality [26]. Virtual reality (VR) has many beneficial effects on society, especially in fields such as education, healthcare, and social engagement. It improves educational experiences with immersive simulations, offers accessible healthcare options, and promotes increased empathy and comprehension among various populations. VR facilitates novel types of social engagement and can enhance inclusivity for individuals with disabilities.

IV. **Reduced Perceived Exertion**

Reduced perceived exertion (RPE) denotes the personal sense of how intensely the body is exerting itself during physical activity, which can be affected by multiple factors and techniques. Fundamentally, it is the degree to which effort is perceived, even if the real physical exertion (such as heart rate or muscle fatigue) may not completely align with that perception. Virtual reality (VR) can diminish the feeling of effort during workouts, possibly making exercise more enjoyable and encouraging. Research indicates that VR can result in reduced perceived exertion (RPE) compared with conventional exercise, despite physiological indicators such as heart rate and oxygen consumption showing comparable or greater effort levels. This phenomenon probably arises from the captivating and immersive qualities of VR, which may divert users' attention from the discomfort of physical activity and enhance their concentration in the virtual space.

8.2.2 AR for Wellness

Augmented reality (AR) has been progressively applied to improve wellness by generating engaging and interactive experiences for fitness, mental well-being, and healthcare. AR can turn workouts into games, create virtual spaces for meditation, and assist in medical training and educating patients. AR can change fitness routines into fun activities, enhancing enjoyment and motivation [27].

I. **Real-Time Guidance and Feedback**

AR applications can superimpose digital content onto the physical world, assisting with activities such as correct exercise techniques or ergonomic modifications for work environments. Augmented reality (AR) is becoming a revolutionary tool that projects digital data onto real-world environments instantaneously. A significant application is its ability to offer immediate guidance and feedback, improving learning, efficiency, and decision-making in various sectors. AR devices (such as smart glasses, headsets, or mobile applications) display instructions, visual indicators, or notifications directly in the user's line of sight. Devices such as sensors and cameras monitor the surroundings and user movements, enabling the system to deliver prompt, context-aware responses [28]. This engagement allows for hands-free, sequential instructions during the execution of intricate activities. Workers can obtain dynamic guidance while operating machinery, minimizing training duration and mistakes. Surgeons can view patient information and detailed surgical instructions while keeping their focus on the patient. Technicians can view visual overlays of assembly guidelines or obtain immediate error notifications while fixing equipment. Students can conduct experiments or simulations with immediate feedback and support from AR systems.

II. **Enhanced Awareness**

AR can improve awareness of the environment and possible dangers, increasing safety and lowering the likelihood of incidents. Wellness initiatives seek to enhance physical, mental, and emotional well-being through the encouragement of healthy practices, mindfulness, and active participation [29, 30]. Augmented reality (AR) enhances real-world settings by superimposing digital data, making it an effective means to increase awareness, foster involvement, and generate immersive experiences in wellness programs. AR devices and applications enable users to view live feedback on heart rate, posture, or stress levels integrated into their surroundings. Posture correction applications can identify misalignments and provide immediate guidance, fostering self-awareness regarding body mechanics. AR can display soothing visuals (such as nature imagery, guided breathing techniques, or meditative animations) within a user's environment. This encourages users to become aware of their levels of physical activity and fitness objectives [31]. Organizations utilize AR to improve wellness training, such as ergonomics workshops, where staff receive immediate feedback on their sitting or standing posture. AR awareness programs emphasize the importance of mental breaks, staying hydrated, and managing stress throughout work hours. Apps with AR capabilities scan food products and show calorie counts, nutritional information, and better alternatives right before the user. This promotes conscious eating and improved food selection. Customized feedback increases people's awareness of their behaviors. Engaging in interactive learning enhances retention and involvement more than conventional approaches do. Immediate corrections stop harmful behaviors (e.g., bad posture, excessive strain). Gamification enhances motivation and encourages sustained engagement in wellness initiatives [32]. Augmented reality greatly boosts awareness of wellness initiatives by transforming unseen health aspects into visible, interactive, and engaging elements. Through posture correction, nutrition monitoring, stress management, or gamifying fitness, AR enhances a more engaging and personalized wellness experience.

III. **Gamified Wellness Programs**

AR can be incorporated into wellness initiatives with gamification features, enhancing their appeal and encouraging employees more effectively. Gamified wellness initiatives use game features—such as points, rewards, challenges, and leaderboards—to encourage healthy practices among staff. When integrated with augmented reality (AR), these applications become increasingly immersive and interactive, turning wellness programs from mundane tasks into captivating experiences. Workers can utilize AR applications/glasses that display virtual coaches, fitness targets, or step-count competitions in the workplace. Walking meetings transform into a virtual treasure in which team members earn rewards by achieving step milestones. AR-driven mindfulness games transform calming settings (such as waterfalls or forests) into workplace environments. Workers earn points for consistent breathing routines, meditation sequences, or relaxation periods, transforming stress management into an enjoyable activity [33]. AR wellness applications monitor hydration, diet, or sleep habits, showcasing achievements via visual badges or AR leaderboards. AR designs collaborative wellness challenges, such as company-wide step competitions or mental wellness adventures, where teams engage in enjoyable, health-oriented missions. Both health and collaboration should be improved. New hires discover wellness policies via AR-enabled wellness games, such as ergonomic posture tasks, or "locate the healthy break area" treasures hunts. Gamified AR converts ordinary wellness tasks into fun experiences. Incentive programs promote ongoing positive habits. Instant AR feedback renders health metrics clear and comprehensible. Team-oriented AR games enhance workplace relationships and share enthusiasm. A healthier, more involved employee enhances a positive organizational culture. Numerous wellness initiatives struggle as participants become disengaged or view the activities as monotonous. Augmented reality (AR) transforms wellness programs by enhancing their immersion, interactivity, and enjoyment, resulting in increased engagement and sustained involvement. AR wellness applications offer real-time feedback during workouts, yoga, or relaxation sessions. Personalized experiences maintain users' motivation and interest [34]. Employees can visualize their progress through 3D dashboards, virtual avatars, or AR milestones, making their advancements concrete and gratifying.

AR facilitates team-oriented wellness challenges where coworkers either compete or join forces in virtual settings, enhancing connections at work. Continuous involvement in wellness programs.

IV. **Increased Engagement**

Engaging and interactive experiences enhance involvement in wellness initiatives, leading to increased participation rates and better results. Augmented reality (AR) can substantially enhance involvement by fostering interactive, tailored, and inspiring experiences. AR can enhance workouts by incorporating gamification, provide immediate feedback to improve performance, and broaden accessibility to a wider audience. This results in increased motivation, improved commitment to wellness objectives, and a more favourable overall experience. AR applications can superimpose virtual components onto the physical environment, turning exercise routines into engaging games filled with challenges, incentives, and tailored feedback. AR can tailor workouts to fit personal fitness levels, preferences, and objectives, enhancing their relevance and motivation. Gamified elements such as points, badges, and leaderboards can increase motivation and motivate users to push their limits [35]. AR workouts are available on smartphones and tablets, allowing those with hectic timetables or living in remote locations to engage more easily. AR can be modified to suit various fitness levels and capabilities, thereby making wellness initiatives more inclusive and reachable for a broader spectrum of people. Additionally, AR offers immediate feedback on technique, advancement, and performance, aiding users in staying on the course and modifying their workouts when necessary [36]. It can help users establish achievable goals and monitor their advancement, promoting a feeling of achievement and encouraging ongoing engagement.

V. **Personalized Experiences**

VR and AR can customize wellness programs according to individual needs and preferences, offering tailored guidance and assistance. Virtual reality (VR) improves customized experiences by generating engaging, interactive settings designed to meet personal demands and tastes. This enables personalized learning, product discovery, and emotional involvement, resulting in a stronger bond with content and brands. VR can develop scenarios that adjust to various learning styles and speeds, enabling students to learn at their own pace and in a manner that fits them best [37]. VR completely engages students, concentrating their senses on the subject matter and potentially enhancing knowledge retention. AI is capable of assessing user actions in real time and modifying VR settings and situations, resulting in customized learning experiences.

Participation is one of the greatest difficulties in wellness initiatives. Numerous employees or participants disengage because of insufficient motivation, repetitive tasks, or minimal understanding of the advantages. Augmented reality (AR), which integrates digital content with the real world, has been shown to enhance wellness programs by making them more interactive, motivating, and tailored—thus increasing engagement. Rather than going through fixed manuals, employees participate in wellness activities through 3D visual enhancements (e.g., live AR feedback for posture adjustment). This enhances wellness sessions by being more enjoyable, engaging, and unforgettable. AR adds points, badges, leaderboards, and challenges to wellness activities. Workers can participate in AR-enabled step-count competitions or hydration contests, transforming healthy behaviors into competitive and engaging activities.

8.3 Digital Wellness Twins and Simulation Models

Digital Wellness Twins utilize simulation models to provide a virtual depiction of a person's health and well-being, enabling tailored health insights, predictive analysis, and preemptive actions. They employ live data from wearable devices, health records, and additional sources to generate a dynamic digital twin, which can subsequently be used to model different scenarios and forecast possible health results. This method allows customized interventions and preventive strategies, ultimately focusing on enhancing personal well-being and streamlining healthcare provision. A digital twin represents a virtual representation of a physical being, specifically focusing on an individual's health and well-being [38]. It is more than a fixed depiction; it is an active, changing framework that mirrors the person's present condition and can receive real-time updates from multiple data sources. Digital twins use simulation models to forecast future health results and assess the possible effects of various interventions. These models allow for the examination of different "what-if" scenarios, including the effects of lifestyle modifications, changes in medication, or even how environmental factors influence a person's health.

8.3.1 Digital Twin Technology

The three key elements of digital twins include data gathering, data representation, and data utilization. The digital twin employs four technologies to gather and archive real-time data, extract information for meaningful insights, and generate a digital replica of a physical item. These technologies encompass the Internet of Things (IoT), artificial intelligence (AI), extended reality (XR), and the cloud (Fig. 8.1). Moreover, digital twins employ specific technologies to varying degrees on the basis of the application type. Internet of Things (IoT): The IoT signifies a vast network of interconnected "things." Relationships exist among objects–objects, individuals–objects, or individuals–individuals. Digital twins rely on the IoT as their main technology across all applications. By 2027, more than 90% of all IoT platforms will feature digital twins. The IoT employs sensors to gather information from physical objects. The information sent by the IoT is utilized to generate a virtual replica of a tangible item. The digital version can subsequently be examined, adjusted, and enhanced. The IoT continuously refreshes information and aids digital twin applications in generating a real-time virtual model of a physical item. Consequently, all digital twin applications utilize the IoT as their main technology. The technology effectively saves and retrieves information via the internet. Cloud computing offers digital twin data processing technology and cloud-based data storage solutions. Cloud computing enables digital twins to store extensive data in the virtual cloud, allowing easy access to necessary information from anywhere. Cloud computing allows digital twins to efficiently diminish the processing time of intricate systems and address the challenges of retaining vast data volumes. Artificial intelligence (AI): As a branch of computer science, AI aims to replicate the foundations of intelligence to develop a novel intelligent machine that can react similarly to human-to-human intelligence. The fields of AI research include robotics, image identification, and language processing. Neural networks, machine learning, deep learning, and expert systems enable AI to support digital twins by offering an advanced analytical tool that automatically assesses acquired data, delivers valuable insights, forecasts outcomes,

Fig. 8.1 Digital twin technologies

and offers recommendations for avoiding potential issues. Extended reality (XR) refers to a broad category encompassing immersive technologies such as virtual reality (VR), augmented reality (AR), and mixed reality (MR). These technologies can integrate the physical and digital realms, enhancing the reality that people perceive. XR generates virtual representations of items that allow digital and physical objects to coexist and engage in real-time.

8.3.2 Applications of Digital Wellness Twins and Simulation Models

Digital wellness twins and simulation models are revolutionizing healthcare and personal well-being by generating virtual representations of individuals who reflect their physical, mental, and behavioral conditions. These models enable ongoing tracking of vital signs, lifestyle behaviors, and environmental influences to forecast health risks and tailor wellness strategies. In professional settings, they can replicate stress levels, exhaustion, and efficiency to create better environments, whereas in medical contexts, they assist in evaluating treatment approaches or recovery programs without endangering patients. Through the incorporation of live data and predictive analysis, digital wellness twins and simulations facilitate proactive measures, improve decision-making, and foster comprehensive well-being customized to each person's distinct requirements.

I. **Personalized Medicine**

Digital twins for individual patients can be developed by combining diverse data, such as biometric data, lifestyle behaviors, and genetic details. This enables customized treatment strategies, precise diagnoses, and individualized care approaches. Digital wellness twins and simulation models are revolutionizing personalized medicine by facilitating highly tailored healthcare approaches [39]. A digital wellness twin is a virtual model of an individual's physiological and behavioral data, which are consistently refreshed with real-time updates from wearable gadgets, electronic health records, and genetic information. Through the integration of these data, simulation models can predict disease advancement, enhance treatment strategies, and evaluate personalized lifestyle interventions for each individual. Healthcare providers may utilize these tools to model the impacts of various medications, diets, or exercise plans prior to application, thus minimizing risks and enhancing results. Moreover, digital wellness twins enable proactive health surveillance, early identification of possible problems, and more accurate modifications to treatments, heralding a new age of data-informed, patient-focused care.

II. **Disease Management**

Digital twins can illustrate disease advancement, emulate treatment outcomes, and assist in creating tailored approaches for handling chronic illnesses. Digital wellness twins and simulation models are transforming disease management by providing sophisticated, data-informed methods to track, anticipate, and address chronic and acute ailments. By constructing a dynamic digital model of a patient with real-time health information—comprising vital signs, laboratory findings, and lifestyle elements—clinicians can accurately simulate disease progression and treatment outcomes [40]. For chronic conditions such as diabetes, heart disease, or asthma, digital wellness twins allow healthcare teams to experiment with different medication plans, behavioral strategies, and lifestyle modifications in a virtual environment prior to implementing them in reality. This method aids in identifying ideal management strategies, avoiding complications, and reducing negative effects. Additionally, simulation models can assist in remote monitoring and prompt intervention by identifying subtle changes that could signal disease worsening, thereby enhancing patient outcomes and lowering hospital admissions. These technologies ultimately promote customized, proactive, and effective disease management designed for each person's distinct health profile.

III. **Surgical Planning and Training**

Patient digital twins can simulate surgeries, strategize intricate operations, and educate surgeons, minimizing risks and enhancing results. Digital wellness twins are personalized, interactive virtual representations of an individual's anatomy and physiology. They are built by combining multimodal data—such as medical imaging (CT, MRI), physiological tracking, electronic health records, and occasionally genomics—into a cohesive, adjustable model. These twins may be static (unchanging 3D representations), functional (mimicking biomechanical reactions), shadow (refchanged in real time throughout surgery), or intelligent (AI-driven, anticipatory, and aiding in decision-making). Simulation models in surgery denote computer-based or virtual settings that mimic surgical situations for preparation, practice, and education. These consist of 3D anatomical representations, physics-driven simulations, VR/AR settings, and tactile feedback systems. By incorporating digital twins, simulation models facilitate preoperative planning, intraoperative direction, and postoperative evaluation.

IV. **Drug Discovery and Development**

Digital twins can model drug interactions, forecast treatment effectiveness, and speed up the drug development process. Digital Wellness Twins and simulation models are progressively changing drug discovery and development by producing highly individualized and predictive virtual representations of human biology. These digital twins combine multiomics data, physiological factors, and disease conditions to model how a person's body may react to a drug candidate, facilitating a more precise prediction of effectiveness and side effects early in the development stage. Simulation models enable in silico evaluation of drug compounds, enhancing molecular design and minimizing the necessity for expensive and time-consuming laboratory tests. Furthermore, digital twins facilitate virtual clinical trials by simulating various patient demographics, enhancing trial design, optimizing patient stratification, and lowering the chances of trial disappointment. This customized, data-informed strategy speeds up drug development schedules, improves safety profiles, and ultimately aids in the generation of more effective, personalized treatments.

V. **Hospital Management**

Digital twins enhance hospital operations by modelling patient movement, resource distribution, and emergency response situations, leading to increased efficiency and better patient outcomes. Digital wellness twins and simulation models are critical in enhancing hospital management by offering real-time, data-informed perspectives on patient movement, resource distribution, and operational effectiveness. By developing virtual replicas of healthcare systems—such as patient admissions, staffing levels, equipment utilization, and bed availability—these models allow managers to simulate different scenarios and predict surges in demand or potential bottlenecks. This enables proactive planning, enhanced readiness for emergencies, and more efficient use of essential resources such as operating theaters and intensive care units. Moreover, incorporating patient-specific digital twins assists in customizing care pathways, lowering the length of stay and readmission rates. Simulation models aid in training hospital personnel in crisis management and protocol compliance, improving overall safety and care quality. Collectively, these technologies enhance informed decision-making, reduce costs, and lead to better patient outcomes in intricate healthcare settings.

VI. **Integration of Wearable Technology**

Digital twins utilize data from wearable devices to enable ongoing monitoring, timely identification of possible health concerns, and preventative measures. The fusion of wearable tech with digital wellness twins and simulation models is transforming hospital management by facilitating ongoing, real-time tracking of patient health and operational performance. Wearable devices gather vital signs, activity metrics, and various physiological information that contributes to digital twins, producing current, individualized patient profiles. This live data integration enables hospitals to anticipate patient decline in advance, enhance bed management, and customize treatment strategies in real time [40]. From an operational perspective, simulation models utilize wearable data to predict patient movement, staff demands, and resource requirements more precisely, improving scheduling and minimizing wait times. Integrating wearable devices with digital twins allows hospitals to increase patient safety, optimize workflows, and facilitate proactive interventions, thereby increasing the quality and efficiency of healthcare services.

VII. **Supply Chain Optimization**

Digital twins are capable of modelling supply chain activities, enhancing inventory control, forecasting demand shifts, and strengthening overall supply chain robustness. Digital wellness twins and simulation models greatly improve supply chain optimization by offering precise, real-time virtual depictions of intricate logistics networks. These models combine information from suppliers, producers, distributors, and inventory systems to forecast and model supply chain performance across different scenarios. In healthcare and pharmaceutical supply chains, digital twins facilitate accurate demand forecasting, aiding in the timely provision of essential medical supplies, medications, and equipment. Simulation models enhance risk assessment by pinpointing possible bottlenecks, disruptions, or resource deficits, enabling stakeholders to proactively create contingency plans. These technologies also aid in inventory optimization by harmonizing stock levels to minimize waste and expenses while ensuring preparedness. In general, digital twins and simulation models enhance efficiency, resilience, and responsiveness in supply chain management, resulting in better patient care and operational sustainability.

VIII. **Product Development and Manufacturing**

Digital twins enable the design, testing, and optimization of products within a virtual setting, minimizing development costs and time. Digital wellness twins and simulation models significantly enhance product development and manufacturing by facilitating virtual prototyping, optimizing processes, and customizing product design. In production, these models replicate workflows and production lines to pinpoint inefficiencies, anticipate equipment maintenance requirements, and minimize downtime, thus improving overall operational effectiveness. In product development, digital twins mimic physical products and their interactions in real-world scenarios, enabling engineers to evaluate performance, durability, and safety without expensive physical prototypes. When utilized in personalized healthcare items—such as medical devices or wearables—digital wellness twins customize design and functionality to meet specific patient requirements, enhancing effectiveness and user satisfaction. This integration speeds up innovation cycles, lowers expenses, reduces waste, and guarantees improved quality and customization in manufacturing processes.

IX. **Urban Planning**

Digital twins are capable of representing urban settings, modelling traffic patterns, and enhancing infrastructure development, resulting in more effective and sustainable urban areas [41]. Digital wellness twins and simulation frameworks are increasingly utilized in urban design to develop intricate virtual models of cities that combine environmental, infrastructure, and public health information. These models allow planners to replicate the effects of urban development initiatives on public health, traffic patterns, pollution rates, and resource usage, facilitating data-informed decision-making. By utilizing live data from sensors and wearables, digital twins assist in tracking community health metrics and forecasting the impact of alterations in urban planning—such as green areas, transit systems, or housing density—on the well-being and quality of life of residents. Simulation models also enhance emergency preparedness by simulating disaster scenarios and improving response strategies. By means of these applications, digital twins aid in creating healthier, more sustainable, and resilient urban settings designed to meet the requirements of their inhabitants.

X. **Disaster Management**

Digital twins can model different disaster situations, assisting emergency responders in preparation, planning, and resource optimization. Digital wellness twins and simulation models serve as essential assets in disaster management, providing detailed virtual depictions of communities, infrastructures, and health systems to improve preparedness, response, and recovery initiatives. By incorporating real-time information from sensors, wearables, and environmental monitoring, digital twins emulate the effects and developments of natural or human-induced disasters such as earthquakes, floods, or pandemics on community health and essential infrastructure. These simulations assist emergency planners in predicting resource requirements, improving evacuation pathways, and aligning medical and logistical responses with increased accuracy. Moreover, digital twins facilitate scenario simulations and training for disaster response teams, enhancing preparedness and decision-making during critical situations. In the end, leveraging digital wellness twins and simulation models facilitates more efficient, prompt, and life-saving responses in emergencies, decreasing fatalities and accelerating community recovery.

Figure 8.2 depicts the idea of a digital twin system, a virtual representation of a physical object utilized for real-time monitoring, analysis, and performance optimization. An analysis of the elements in the diagram is as follows: (i) Physical entity (top center): Symbolizes the actual object, system, or person (depicted as a human figure in this case). This refers to the real topic under investigation or reflected in the digital realm. (ii) Virtual entity (bottom center): Symbolizes the digital representation of the physical entity. It is created with information gathered from the physical entity and mimics its actions, conditions, and processes. (iii) Data (Left side): Information gathered from the physical entity (e.g., sensors, monitoring systems, reports). This information is constantly supplied to the virtual entity to maintain its alignment with the real world. (iv) Digital Twin Aggregation (Top-Left): This involves merging various data sources and layers to develop a comprehensive and precise digital twin. It collects information from various systems and devices to increase the

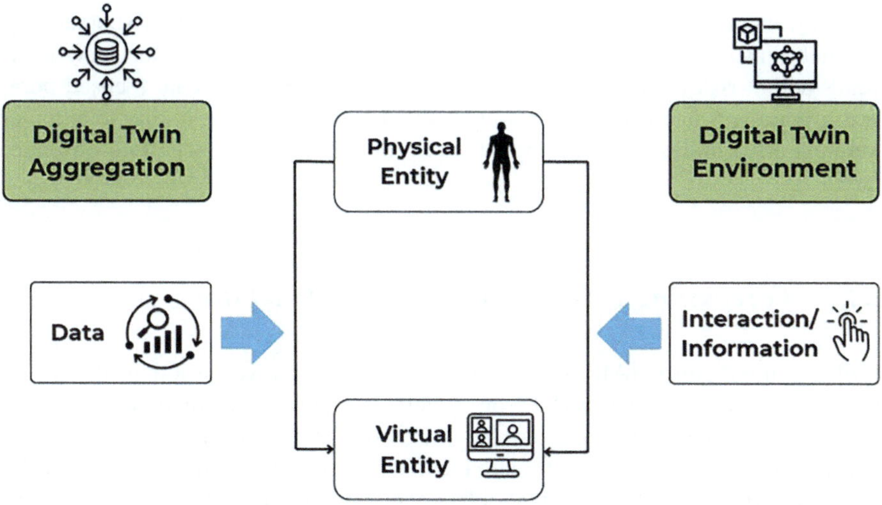

Fig. 8.2 Digital twin system (concept)

precision of the virtual model. Digital Twin Environment (Top-Right): The framework in which digital twins function, communicate, and connect with additional systems. This consists of cloud infrastructures, IoT frameworks, and AI settings that enhance the digital twin's capabilities. (v) Interaction/Data (Right side): The bidirectional communication link. Data circulate in both directions between the digital twin setting and its users (decision-makers, operators, or systems). Users can engage with the digital twin to evaluate scenarios, foresee results, and enhance efficiency.

8.4 Predictive AI for Preventive and Personalized Wellness

Predictive AI is transforming preventive and personalized wellness through early disease identification, customized treatment strategies, and proactive health care management. AI algorithms assess extensive health information, such as genetic data, medical records, lifestyle data, and live biometric data from wearable devices, to forecast possible health risks and customize interventions for individuals. This transition from reactive to proactive care aids individuals and healthcare providers in making informed choices, enhancing treatment approaches, and ultimately improving overall health results. AI algorithms can evaluate medical images, such as X-rays and MRIs, to detect slight indications of illness in their initial phases, possibly prior to any symptoms manifesting. Wearable devices and medical sensors consistently track vital signs, enabling the immediate identification of irregularities that may suggest health problems [42]. Through the examination of electronic medical records and laboratory results, AI can recognize trends that could suggest a tendency toward specific illnesses. AI can examine extensive datasets from different sources (e.g., hospitals, insurance firms, and public health organizations) to detect new health trends and at-risk populations. These data can be utilized to create focused public health initiatives and distribute resources efficiently. AI has the ability to automate administrative duties, allowing healthcare professionals to concentrate on patient care.

8.4.1 Forecasting Health Issues in Predictive Care

Artificial intelligence (AI) is ushering in a transformative period in the healthcare sector, changing the model from a uniform approach to medicine to a more detailed, predictive, individualized, and preventive comprehensive care framework. This change is not merely a futuristic idea; it is currently altering the ways healthcare providers, patients, and stakeholders connect, handle, and comprehend health and wellness. As the volume of data from genomics, electronic health records, and wearable devices continues to rise, AI's capacity to analyse and learn from intricate datasets creates unmatched possibilities in the diagnosis, treatment, and prevention of diseases. It is setting the stage for personalized medicine, where therapies and interventions are customized to meet individual patients' requirements, preferences, and genetic profiles. Additionally, the foresight provided by AI allows healthcare workers to intervene before a condition worsens, thus improving results and lowering expenses. This study investigated the effects of AI on healthcare and its ability to revolutionize the medical field into a more efficient, compassionate, and tailored system while also addressing the obstacles that need to be addressed to facilitate this new phase of medicine.

The arrival of predictive care signifies an important advancement in proactive medical treatment [43]. For example, algorithms driven by AI and models utilizing machine learning enable healthcare professionals to effortlessly assess large volumes of historical and real-time information. This evaluation encompasses demographics, medical histories, genetic data, and social factors affecting health. Doctors and medical professionals can effectively predict possible health problems in patients by recognizing trends and relationships within these datasets. Predictive analytics can flag the risks of chronic conditions such as diabetes, heart disease, or cancer at early stages, enabling timely intervention and potentially saving lives. Personalized medicine, or precision medicine, is a burgeoning field that epitomizes the fusion of AI and healthcare. Through detailed analysis of a patient's genetic makeup, lifestyle, and environmental factors, medical practitioners can now design treatment plans that are specifically tailored to the individual. This granularity in care leads to increased treatment effectiveness and minimizes adverse effects. For example, oncologists can utilize AI-driven genomic sequencing to identify the exact type and mutation of cancer cells, subsequently prescribing targeted therapies that directly attack cancerous cells without harming healthy tissues [44]. Personalized care also extends to creating unique wellness plans, emphasizing nutrition, exercise, and mental well-being, customized to an individual's needs and preferences.

8.4.2 The Power of Machine Learning Algorithms

Machine learning, a branch of AI, is transforming healthcare by improving diagnostics, predictive analytics, and tailored medicine. It utilizes algorithms capable of rapidly analysing intricate and vast amounts of clinical data to uncover patterns, correlations, and trends. Possible uses of this technology include the following: (i) Personalized medicine: Machine learning aids in forecasting the effectiveness of treatment strategies customized for individual patient requirements. Its effectiveness in healthcare is evident in supervised learning, where known outcomes are employed to examine medical images and clinical information. (ii) Identifying new therapies: Machine learning is being used to explore patient records and identify subtle variations in vital signs, leading to the development of innovative therapies and enabling more personalized treatments. (iii) Substituting rule-based systems: Conventional rule-based systems in healthcare are slowly being supplanted by machine learning techniques that can analyse data through specialized medical algorithms without the constraints associated with extensive rule sets.

The accompanying Fig. 8.3 depicts the various sources that provide healthcare data, emphasizing the interlinked ecosystem that underpins contemporary healthcare systems. At the core is healthcare data, which are enhanced by various inputs: genomic data (genetic details that support precision medicine), medical images (such as X-rays, MRIs, and CT scans for diagnosis), and medical records (patient history, prescriptions, and treatment information). Moreover, environmental data (elements such as air quality, climate, and pollution) and the internet (internet-based health platforms and digital health services) offer essential contexts for comprehending patient health. Contemporary technology enhances this data reservoir via social media (patient insights and health patterns), wearable devices (instant monitoring of vital signs), and mobile phone data (applications and communication tools for health monitoring). Together, these resources combine to form an all-encompassing, data-informed healthcare system that facilitates tailored treatment, anticipatory analytics, and enhanced decision-making for improved patient results.

8.4.3 Understanding the Role of Natural Language Processors

The increase in the use of deep learning in healthcare has involved its use in speech recognition, employing natural language processing, or NLP, a branch of machine learning. As this technology advances, healthcare professionals need to adjust to

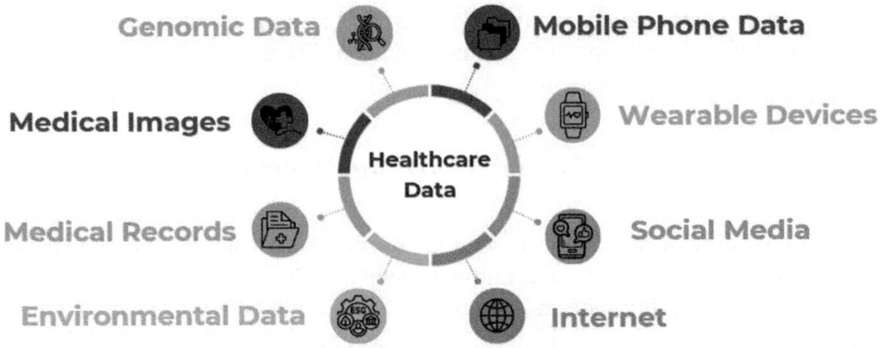

Fig. 8.3 Healthcare predictive analytics using machine learning

comprehend and effectively apply deep learning models in their work [45]. The incorporation of NLP into healthcare has introduced an advanced method for interpreting and using human language, with implementations of the following: (i) Diagnosis and Treatment Customization: NLP has played a crucial role in precisely diagnosing disease by extracting essential information from patients' health records, including examinations and results, thereby further tailoring their treatments; and (ii) Improving clinical procedures: NLP has significantly enhanced healthcare operations by efficiently handling vast amounts of complicated data, providing more personalized, timely service for patients. The dependence on algorithms also raises concerns regarding accountability and transparency, especially when AI-generated choices result in unforeseen or unfavourable results. Together with establishing a system where AI acts as an aid instead of a substitute for human expertise, fostering trust between healthcare professionals and patients will be essential in addressing these obstacles.

8.4.4 Rise of Telemedicine and Virtual Health Assistants

Telemedicine and virtual health aides represent another domain in which AI is advancing within healthcare. As the demand for remote medical consultations grows, especially following global health emergencies such as the COVID-19 pandemic, AI-driven chatbots and virtual health platforms are offering prompt and accessible healthcare guidance. These virtual helpers can manage standard questions, direct patients through symptoms, and provide mental health assistance [46, 47]. These AI-powered systems guarantee that high-quality healthcare goes beyond the limits of conventional medical institutions, serving underserved and isolated communities by supporting human medical practitioners. The emergence of telemedicine and virtual health aides signifies a significant change in contemporary healthcare, propelled by progress in digital technologies and the increasing demand for accessible, effective medical services. Telemedicine allows patients to engage with healthcare providers remotely via video calls, mobile applications, and online platforms, overcoming geographical limitations and lessening pressure on physical medical facilities. In addition, AI-driven virtual health assistants provide tailored help by responding to medical questions, tracking vital signs, providing medication alerts, and assisting patients with chronic condition management. Collectively, these advancements improve convenience and affordability while also providing patients with ongoing, proactive care, fundamentally transforming health care delivery into a more patient-focused, data-informed, and inclusive framework.

The advancement of AI for comprehensive wellness indicates transformative changes in how people and organizations view health, well-being, and personal development [48]. Owing to advancements in predictive analytics, tailored recommendations, digital twins, and immersive technologies such as AR/VR, AI will transition from reactive healthcare to proactive and preventive wellness strategies. Through the harmonious integration of physical, mental, social, and occupational health, AI promotes more balanced living, lowers stress levels, and improves overall quality of life. To maximize the potential of these innovations, it is essential that ethical factors such as data privacy, inclusivity, and transparency take precedence [49]. In the end, AI-based comprehensive wellness enables people to manage their health while allowing organizations to foster healthier, more resilient, and sustainable settings.

8.5 Conclusions

The future of AI in holistic wellness is set to deliver groundbreaking improvements by effortlessly merging physical, mental, and social health aspects into tailored, adaptable care solutions. New AI technologies will utilize extensive and varied data sources—from genomics and wearable devices to behavioral and environmental aspects—to develop more advanced digital wellness twins that facilitate proactive health management and early intervention. Progress in explainable AI, federated learning, and ethical guidelines will bolster trust, privacy, and inclusivity, making sure that AI-based wellness tools are reachable and fair. Additionally, virtual coaches driven by AI, continuous monitoring, and predictive analytics will enable people to make educated lifestyle decisions, enhance mental health, and cultivate resilient communities. As these advancements progress, AI will be pivotal in transforming healthcare from reactive interventions to a consistent, personalized, and comprehensive wellness model that enhances quality of life worldwide.

References

1. Dey, D. N. C. (2024). Harmonizing the threads of well-being: AI (artificial intelligence), spirituality, healthcare, and management. *Spirituality, Healthcare, and Management, 14*, 131.

2. Fabbrizio, A., Fucarino, A., Cantoia, M., De Giorgio, A., Garrido, N. D., Iuliano, E., et al. (2023). Smart devices for health and wellness applied to tele-exercise: an overview of new trends and technologies such as IoT and AI. In *Healthcare* (p. 1805). MDPI.
3. Ng, J. Y., Cramer, H., Lee, M. S., & Moher, D. (2024). Traditional, complementary, and integrative medicine and artificial intelligence: Novel opportunities in healthcare. *Integrative Medicine Research, 13*(1), 101024.
4. Olawade, D. B., Wada, O. Z., Odetayo, A., David-Olawade, A. C., Asaolu, F., & Eberhardt, J. (2024). Enhancing mental health with artificial intelligence: Current trends and future prospects. *Journal of Medicine, Surgery, and Public Health, 3*, 100099.
5. Zhao, Y., & Sun, P. (2025). Artificial intelligence in the promotion of human well-being: Current trends and future directions. *Well-Being Sciences Review, 1*(1), 3–8
6. Karthikeyan, V., & Visu, Y. P. (2025). 14 future directions trends in digital and health and wellness. *Cybersecurity in Healthcare Applications, 246*, 3254.
7. Dhinakaran, D., Raja, S. E., Jasmine, J. J., Kumar, P. V., & Ramani, R. (2025). The future of well-being: AI-powered health management with privacy at its core. *Wellness Management Powered by AI Technologies, 64*, 363–402.
8. Sabharwal, P., & Singh, I. (2024). Anticipating tomorrow: Future trends and predictions in Ayurveda entrepreneurship. *Future Health, 34*, 1–6.
9. Qirtas, M. M., Zafeiridi, E., White, E. B., & Pesch, D. (2024). Evolving AI for wellness: Dynamic and personalized real-time loneliness detection using passive sensing.
10. Zhang, L., & Shi, M. (2025). Unlocking the potential of AI-driven tourism for healing and well-being: A new frontier in digital health. In *Proceedings of the 2025 11th international symposium on system security, safety, and reliability (ISSSR)* (pp. 441–449). IEEE.
11. Tiwari, M., & Waoo, A. A. (2024). Transforming healthcare: The synergistic fusion of AI and IoT for intelligent, personalized well-being. In *Revolutionizing healthcare: AI integration with IoT for enhanced patient outcomes* (pp. 109–149). Springer.
12. Rosca, C. M., & Stancu, A. (2024). Fusing machine learning and AI to create a framework for employee well-being in the era of industry 5.0. *Applied Sciences, 14*(23), 2076–3417.
13. Rao, T. V. N., Patel, G. S., Bhavana, S., & Swetha, C. (2025). AI-driven decision support system for workplace wellbeing. In *Insights on wellbeing and happiness in the workplace* (pp. 1–18). IGI Global Scientific Publishing.
14. Sundarajan, A. (2025). Enhancing workplace productivity and well-being using AI agent.
15. Mary, B. J. (2018) *Predictive analytics for employee wellness: An AI-driven solution to minimize workplace burnout*. Britney Johnson Mary New York University United States.
16. Mittal, S., Wijethilak, H. P., & Chandel, A. (2025). Effectiveness of AI-powered wellness in promoting employee wellbeing and happiness. In *Insights on wellbeing and happiness in the workplace* (pp. 87–108). IGI Global Scientific Publishing.
17. Indriya, S., Fazlurrahman, H., & Fitro, A. (2024). Machine learning-based prediction of the impact of mental health policies on employee productivity. In *International conference on digital business innovation and technology management (ICONBIT)*.
18. Dikshit, S., Grover, Y., Shukla, P., Mishra, A., Sahu, Y., Kumar, C., & Gupta, M. (2024). Empowering employee wellness and building resilience in demanding work settings through predictive analytics. *EAI Endorsed Transactions on Internet of Things, 10*, 4542.
19. Piras, G., Agostinelli, S., & Muzi, F. (2025). Smart buildings and digital twin to monitoring the efficiency and wellness of working environments: A case study on IoT integration and data-driven management. *Applied Sciences, 15*(9), 4939.
20. Albraikan, A. (2019). *InHarmony: A digital twin for emotional well-being* (Doctoral dissertation, Université d'Ottawa/University of Ottawa).
21. Hii, D. J. C., & Hasama, T. (2024). Towards the digital twinning and simulation of a smart building for well-being. In *Proceedings of the 2024 winter simulation conference (WSC)* (pp. 726–737). IEEE.
22. Gnanamalar, R. H. (2024). Human digital twin processes and their future. In *Transforming industry using digital twin technology* (pp. 187–217). Springer.

23. Gupta, P., Sharma, M., & Joshi, S. (2025). Digital health: Navigating digital transformation and digital stress. In *Sustainable digitalization strategies in business and healthcare* (pp. 261–294). IGI Global Scientific Publishing.
24. Lindstrom, E. (2023). *Feasibility of a human capital digital twin.* Massachusetts Institute of Technology.
25. Pietrantoni, L., San Román-Niaves, M., & De Angelis, M. (2024). Human trust and digital twins in a human factors and ergonomic framework. In *Proceedings of the 2024 IEEE 29th international conference on emerging technologies and factory automation (ETFA)* (pp. 1–9). IEEE.
26. Battini, D., Berti, N., Cella, C., Faroni, M., Garza, P., Guidolin, M., et al. (2024). Designing the operator of the future: The architecture of human digital twin systems. *IFAC-PapersOnLine, 58*(19), 355–360.
27. Tan, C. H., Koo, A. C., Rahmat, H., Siew, W. F., Cheang, A. W. O., & Sharji, E. A. (2023). Workplace wellness, mental health literacy, and usage intention of e-mental health amongst digital workers during the COVID-19 pandemic. *International Journal of Mental Health Promotion, 25*(1), 99–126.
28. Sujanthi, S., VB, A. R., Poovizhi, M., Kayathri, T. L., Balaji, P., & Hariharan, M. (2025). Smart wellness: AI-driven IoT and wearable sensors for enhanced workplace well-being. In *Proceedings of the 2025 international conference on multi-agent systems for collaborative intelligence (ICMSCI)* (pp. 502–509). IEEE.
29. Cole, A. (2023). The ethics of the personal digital twin. In *Human data interaction, disadvantage and skills in the community: Enabling cross-sector environments for postdigital inclusion* (pp. 79–92). Springer.
30. Khanfar, N. M., Nafei, W. A., & Mujtaba, B. G. (2024). Impact of digital twinning on health service quality and sustainable competitive advantage: A study with private hospital professionals. *Health Economics and Management Review, 5*(4), 75–90.
31. Mikołajewski, D., Piszcz, A., Rojek, I., & Galas, K. (2024). Machine learning supporting virtual reality and brain–computer interface to assess work–life balance conditions for employees. *Electronics, 13*(22), 4489.
32. Pabolu, V. K. R., & Shrivastava, D. (2021). A dynamic job rotation scheduling conceptual framework by a human representing digital twin. *Procedia CIRP, 104*, 1367–1372.
33. Sirigu, G., Carminati, B., & Ferrari, E. (2022, December). Privacy and security issues for human digital twins. In *Proceedings of the 2022 IEEE 4th international conference on trust, privacy and security in intelligent systems, and applications (TPS-ISA)* (pp. 1–9). IEEE.
34. Davila-Gonzalez, S., & Martin, S. (2024). Human digital twin in industry 5.0: A holistic approach to worker safety and well-being through advanced AI and emotional analytics. *Sensors, 24*(2), 655.
35. Lektauers, A., Pecerska, J., Bolsakovs, V., Romanovs, A., Grabis, J., & Teilans, A. (2021). A multimodel approach for simulation-based digital twin in resilient services. *WSEAS Transmission System Control, 16*, 133–145.
36. Ogunseiju, O. R., Olayiwola, J., Akanmu, A. A., & Nnaji, C. (2021). Digital twin-driven framework for improving self-management of ergonomic risks. *Smart and Sustainable Built Environment, 10*(3), 403–419.
37. Barat, S., Mulpuru, D., Yadav, A., Basu, A., Kulkarni, V., Samudrala, S., et al. (2022). A digital twin based approach for ensuring business continuity plan and safe return to workplace. In *Proceedings of the 2022 annual modelling and simulation conference (ANNSIM)* (pp. 126–139). IEEE.
38. He, Q., Li, L., Li, D., Peng, T., Zhang, X., Cai, Y., et al. (2024). From digital human modelling to human digital twin: Framework and perspectives in human factors. *Chinese Journal of Mechanical Engineering, 37*(1), 9.
39. Pollitz, K., & Rae, M. (2020). Trends in workplace wellness programs and evolving federal standards. *KFF*. https://www.kff.org/private-insurance/issue-brief/trends-inworkplace-wellness-programs-and-evolving-federal-standards

40. Sopow, E. (2020). Aligning workplace wellness with global change: An integrated model. *Journal of Organizational Change Management, 33*(5), 909–923.
41. Arakawa, Y. (2019). Sensing and changing human behavior for workplace wellness. *Journal of Information Processing, 27*, 614–623.
42. Mattke, S., Schnyer, C., & Van Busum, K. R. (2013). A review of the US workplace wellness market. *Rand Health Quarterly, 2*(4), 7.
43. Cahalin, L. P., Myers, J., Kaminsky, L., Briggs, P., Forman, D. E., Patel, M. J., et al. (2014). Current trends in reducing cardiovascular risk factors in the United States: Focus on worksite health and wellness. *Progress in Cardiovascular Diseases, 56*(5), 476–483.
44. Bensimon, P. (2010). *Wellness at work: A choice for a better future.* Correctional Service of Canada, Research Branch.
45. Fathima, F., Muppidi, S., & Manoharan, G. (2025). Workplace wellness: A goal for employee satisfaction-understanding workplace well-being and happiness. In *Development of self-awareness and wellbeing: Global learning challenges in a shifting society* (pp. 541–562). IGI Global Scientific Publishing.
46. Richardson, K. M. (2017). Managing employee stress and wellness in the new millennium. *Journal of Occupational Health Psychology, 22*(3), 423.
47. Alexander, R., & Fry, A. (2019). Strategic design and the future of work-and-wellness. *Design for Health, 3*(1), 135–154.
48. Quintero Vergara, H., & Castro Pineda, A. (2022). A bibliometric review on workplace health and wellness programs: Main research perspectives. *Económicas CUC, 43*(2), 4985.
49. DeVries, G. T., III. (2010). Innovations in workplace wellness: Six new tools to enhance programs and maximize employee health and productivity. *Compensation and Benefits Review, 42*(1), 46–51.

Chapter 9
Case Studies and Global Best Practices

Global wellness best practices encompass a comprehensive approach that extends beyond physical health, emphasizing mental, emotional, social, and financial wellness. Case studies demonstrate the effectiveness of well-rounded programs that incorporate elements such as onsite exercise, mental health support, and financial literacy training. Effective programs utilize technology for tailored experiences, provide adaptable choices to accommodate varied requirements, and cultivate a well-being culture through company policies and staff involvement. Expanding from physical fitness to include mental, emotional, social, and financial wellness. Employing online platforms for tailored wellness programs, health monitoring, and readily available resources. A range of wellness choices, including onsite classes, gym memberships, and online programs, are provided to meet the varied needs and preferences of employees. Initiatives that avert disease and encourage healthy living, such as programs for quitting smoking, managing stress, and controlling weight, are needed. Creating organizational guidelines and fostering a positive workplace culture that promotes self-care, alleviates stress, and enhances overall well-being.

9.1 Case Studies on AI-Driven Well-Being and Data Analysis

I. Tech in Asia—Mental Wellbeing App

To help employees stay healthy, *Intellect* developed a mental health app for Tech in Asia, one of the largest media organizations in Asia. The app offers one-on-one behavioural health coaching, skill-building programmes, and guided stress reduction. The active involvement of senior leaders was crucial to their success because they set an example by using the app and providing feedback, which promoted employee engagement and fostered trust. The average number of coaching sessions per user

S. Majumder and N. Dey, *AI-Driven Wellbeing to Enhance Lives and Work Environments*, Studies in Computational Intelligence 1257,
https://doi.org/10.1007/978-3-032-15735-5_9

increased by 87.5% within eight months of the app's launch; 75% of employees were actively using it, and 55% of them were interacting with professional coaches. While the initiative demonstrated strong outcomes in improving wellbeing, the company recognized that maintaining long-term engagement would require ongoing effort and that data privacy and confidentiality were crucial for employee trust. Even though the program showed promising results in enhancing wellbeing, the business understood that sustained involvement would need constant work and that employee trust would depend on data privacy and confidentiality [1, 2].

II. ***Emotional AI*at a Finnish Research Institute**

To investigate how artificial intelligence can track or infer emotional states via sensors and feedback, a Finnish research institute piloted the use of "*Emotional AI*" with 11 employees. Concerns around privacy, transparency, and ownership of sensitive emotional data surfaced, although participants tended to react favourably when they observed clear benefits such as stress management tools. Research has emphasized the need for transparency and voluntary engagement in AI-driven well-being programs, highlighting that while strong policies and clear communication help foster some trust, the perception of surveillance can be just as important as the technology itself [3, 4].

III. **TILES-2018 Dataset/Hospital Workers Study (U.S.)**

To gather comprehensive multimodal data on physiological and behavioural patterns, the TILES-2018 project in the United States followed 212 hospital employees over time via wearable technology, ambient sensors, and surveys. Linking these signals to interactions at work, job performance, and well-being was the goal. The results indicated that ongoing observations may yield important information about stress, sleep, and productivity, opening the door for predictive well-being models. The high resource requirements of data collection, participant adherence to wearable technology, and the problem of converting predictive signals into workable treatments for practical applications are obstacles [5, 6].

IV. **Health Navigator (UK, NHS Collaboration)**

The health navigator used predictive analytics to identify patients who were most at risk of unplanned hospital admissions and offered them proactive health coaching in collaboration with the UK's National Health Service (NHS) and Clinical Commissioning Groups. This program demonstrated the value of early identification and focused support by reducing emergency visits by 36% and unplanned admissions by 25%. Despite these encouraging outcomes, growing the program requires ongoing funding, rigorous control of predictive accuracy to prevent patient misclassification, and resolution of ethical issues about the order of priority for therapies [7].

Together, these case studies show that although AI and data-driven wellbeing programs can greatly enhance productivity, contentment, and health, their effectiveness is mostly dependent on involvement and trust rather than technology. While transparency and robust data governance, as shown in the Finnish Emotion AI study, are essential to resolving concerns about privacy and surveillance, leadership participation, as demonstrated at Tech in Asia, increases adoption by providing a positive example. The TILES-2018 hospital worker research demonstrated that while multimodal data can provide valuable insights into performance and stress, it might be difficult to translate those findings into effective interventions [8]. Finally, the NHS Health Navigator programme highlights the real advantages of predictive analytics in lowering hospital admissions while simultaneously emphasizing the significance of precision, finance, and ethical issues in expanding such initiatives. All of these examples point to the success of AI-driven well-being when it combines strong analytics with human-centered design, transparent communication, and sustained dedication.

9.2 Policy-Driven AI Wellness Initiatives

AI wellness initiatives driven by policy aim to incorporate artificial intelligence to enhance and tailor employee health and well-being initiatives. These efforts utilize AI's potential to improve different facets of wellness, such as physical, mental, and financial health, while prioritizing data privacy and security. Algorithms examine personal data (e.g., health metrics from wearables, lifestyle habits) to develop customized fitness programs, nutrition advice, and stress reduction techniques. AI can detect possible health issues (e.g., burnout, chronic conditions) early via predictive analytics, allowing prompt interventions and preventive actions. Chatbots and platforms powered by AI deliver tailored mental health assistance, granting access to resources, guidance, and therapy. AI can aid in financial planning, budgeting, and establishing goals, enabling employees to handle their finances more efficiently. Strong measures are essential to protect sensitive employee health data during the implementation of AI-based wellness initiatives [9]. AI can increase the availability of wellness resources for employees, regardless of their location or timetable, by offering round-the-clock support and tailored advice. AI Wellness Initiatives Driven by Policy include the following.

9.2.1 Government and National-Level Policies

I. **EU AI Act (2024)**

It encompasses ethical standards and transparency obligations for AI systems in terms of human resources and wellness oversight. Guarantees that wellness tools powered by AI (such as stress detection and virtual mental health aides) uphold privacy and avoid discrimination. Require "human supervision" in AI-driven decision-making that impacts employees. The EU AI Act (2024) represents the first extensive legal framework globally governing artificial intelligence, aimed at guaranteeing that AI systems utilized in various sectors are secure, clear, and consistent with fundamental rights. It classifies AI applications by risk level—unacceptable, high, limited, and minimal with tools for workplace wellness, such as biometric tracking or mental health analytics, which are typically grouped in the "high-risk" category. Under the Act, employers need to ensure transparency, secure informed consent from employees, and uphold human oversight when implementing AI-based wellness solutions. The law additionally highlights equity and non-discrimination, ensuring that sensitive health information is not misused in job-related choices. Through the implementation of these regulations, the EU AI Act safeguards workers' rights while fostering responsible innovation, establishing an international standard for incorporating ethical AI in employee wellness initiatives.

II. **India and the National Digital Health Mission (NDHM)**

India and the National Digital Health Mission establish uniform health IDs and AI-driven health records. Employers can incorporate wellness initiatives (telehealth, predictive health risk evaluations) into regulated structures. The Government of India initiated the National Digital Health Mission (NDHM) as a major program to establish a cohesive digital health framework that enhances the accessibility, efficiency, and transparency of healthcare services. At the core of the mission is the creation of a distinct health ID for each citizen, facilitating secure digital storage and effortless sharing of health records among hospitals, clinics, insurers, and wellness providers. For workplace wellness, NDHM offers a solid framework by permitting AI-driven platforms to combine employee health information with permission, facilitating preventive care, early risk identification, and tailored wellness initiatives. Through the standardisation of data governance, safeguarding privacy, and enhancing interoperability, NDHM enables individuals to manage their health effectively while assisting organisations in providing organized, technology-focused wellness programs within a reliable and regulated environment.

III. **U.S. Corporate Wellness Policies**

Under the Affordable Care Act (ACA) and HIPAA, employer wellness programs utilizing AI-based biometric monitoring must ensure privacy. The Equal Employment Opportunity Commission (EEOC) oversees AI-related health data to avoid bias in employment benefits. U.S. corporate wellness initiatives are shaped by a blend of federal laws, ethical principles, and organizational best practices designed to increase employee well-being while protecting personal rights. The Affordable Care Act (ACA) and HIPAA promote the establishment of wellness programs by employers, requiring them to guarantee the confidentiality of employee health information and prohibit its use for discriminatory actions. The Equal Employment Opportunity Commission (EEOC) also oversees incentive systems to guarantee that involvement in wellness programs is voluntary and fair. Numerous U.S. businesses incorporate AI-driven tools such as biometric assessments, mental health services, and predictive health analytics into these policy structures, merging innovation with regulation.

9.2.2 *Organizational AI Wellness Policies*

I. **Data Privacy and Ethics in Wellness AI**

Employers implement regulations that restrict the ways in which AI gathers and interprets employee health information (e.g., fitness trackers and mental health chatbots). Internally applied GDPR-like principles: consent, transparency, and minimal data gathering. The privacy of data and ethical considerations in wellness AI are essential for preserving employee trust and organizational integrity while employing technology to increase well-being. Given that wellness programs frequently include sensitive health information from wearables, surveys, or AI-driven mental health applications, explicit policies should outline the collection, storage, and use of these data. Ethical guidelines highlight the importance of transparency, obtaining employee consent, limiting data collection, and rigorously safeguarding confidentiality to avert the misuse of health information for discriminatory actions such as hiring or promotions. The incorporation of fairness, accountability, and human oversight into AI-led wellness programs enables organizations to protect individual rights while utilizing data to provide tailored assistance, minimize risks, and promote a healthier, more responsible workplace environment.

II. **AI for Fair Access and Inclusivity**

Policies guarantee that AI-based wellness platforms (such as mental health applications and ergonomic risk assessment tools) are available for a variety of employee demographics, including individuals with disabilities. AI-generated captions and language translation in wellness workshops foster inclusivity and equitable access to workplace wellness, ensuring that health and wellness programs cater to the varied needs of every employee, irrespective of their background, ability, or work situation. By utilizing AI-enhanced translation, speech recognition, and accessibility tools, wellness platforms can eliminate language and communication obstacles, ensuring that resources are accessible to diverse and multilingual teams [10–12]. AI aids in detecting participation gaps by examining demographic and usage information, enabling organizations to customize programs for underrepresented groups such as frontline staff, employees with disabilities, or remote workers. By offering tailored suggestions and fair resource allocation, AI promotes a more inclusive environment where all employees have the same chance to participate in wellness initiatives and take advantage of organizational assistance.

III. **AI-Powered Early Detection Policies**

Employers establish screening procedures where AI detects early signs of burnout, depression, or cardiovascular issues on the basis of patterns (attendance, sleep, workload). Policies specify that intervention occurs through voluntary counselling support rather than punitive actions. AI-driven early detection strategies emphasize the use of sophisticated analytics and machine learning technologies to pinpoint possible health and wellness threats among employees before they develop major problems. Through the examination of data from wearable devices, work habits, absenteeism patterns, and self-reported wellness information, AI can identify early signs of stress, burnout, fatigue, or long-term health issues. These policies guarantee that interventions—such as counselling assistance, health coaching, or workload modifications—are preventative, confidential, and nonpunitive, motivating employees to request help without the concern of stigma. Timely identification safeguards employee health while also lowering long-term medical expenses, decreasing absentee rates, and fostering a more robust, resilient workforce.

9.2.3 Initial Sector-Specific AI Wellness

I. **Healthcare Sector**

Policies promote AI-based wellness dashboards for hospital employees to monitor fatigue, shift stress, and connections to patient safety. Hospitals and clinics employ AI-powered tools to detect indicators of stress and burnout in nurses and physicians by examining work schedules, patient workloads, and electronic health record (EHR) interactions. The Mayo Clinic tested AI-enhanced resilience and stress management programs (SMART) to identify and alleviate nurse burnout [13]. The Mayo Clinic and NHS are testing policies in which AI monitors nurse burnout and proposes automatic shift changes. These platforms evaluate biometric screenings, lifestyle information, and health records of healthcare workers to predict risks for chronic conditions such as diabetes, hypertension, and heart disease. In the healthcare field, AI-powered wellness programs are increasingly utilized to address the specific challenges encountered by medical professionals, including exhaustion, elevated stress levels, and extensive working hours. Hospitals and healthcare systems are utilizing AI to analyse workload trends, forecast health hazards, and offer tailored assistance via mental health chatbots, online counselling, and resilience training services [14]. AI-powered dashboards and predictive analytics assist in detecting initial indications of fatigue or long-term illness among employees, whereas smart scheduling systems enhance shift distributions to minimize overexposure. The incorporation of these tools into everyday activities allows healthcare organizations to not only increase employee wellness but also increase patient safety, decrease absenteeism, and cultivate a nurturing workplace environment.

II. **Tech Sector**

Firms such as Microsoft, Google, and IBM implement internal measures for "Responsible AI Wellness," merging usage oversight (screen fatigue, workload equilibrium) with employee rights to opt out. Within the tech industry, AI wellness programs aim to address digital fatigue, stress linked to elevated workloads, and mental health issues arising from rapidly evolving work environments. Organizations utilize AI-driven tools to measure screen time, oversee work–life balance, and offer tailored prompts that promote breaks, mindfulness, or ergonomic changes. Virtual wellness assistants and chatbots provide private mental health assistance, whereas AI-powered analytics detect initial signs of burnout or disengagement by examining communication trends, workload patterns, and employee input.

9.2.4 Policy Benefits

I. **Trust and Participation**

Workers are more inclined to participate in AI wellness initiatives when robust policies safeguard them. Trust is the cornerstone of an effective workplace wellness policy, as workers are more inclined to participate when their information and health are managed with openness, equity, and privacy [15]. A well-articulated policy that safeguards privacy, fosters inclusivity, and highlights voluntary involvement motivates employees to engage without concern for stigma or exploitation. As trust develops, engagement levels rise, resulting in better health results, lower stress, and enhanced morale throughout the workforce. This establishes a beneficial cycle in which increased involvement enhances confidence in the organization's dedication to employee welfare, ultimately bolstering organizational culture, productivity, and loyalty.

II. **Risk Reduction**

Lowers the likelihood of the inappropriate use of confidential health information. An effectively crafted workplace wellness policy is essential for minimizing risks for individuals and the organization as a whole [16]. Through the promotion of preventive health measures, routine screenings, stress reduction techniques, and lifestyle changes, these policies assist employees in recognizing and managing health concerns before they develop severe problems. This reduces the likelihood of chronic diseases and exhaustion while also decreasing workplace injuries and absenteeism. For employers, wellness policies minimize the financial risk linked to increasing healthcare expenses and insurance claims and decreasing productivity. By actively addressing health and safety risks, organizations build a more resilient workforce and protect long-term organizational viability.

III. **Cultural Integration**

Wellness is incorporated into corporate governance rather than viewed as a discretionary HR perk [17]. Cultural integration in wellness policies guarantees that health and well-being initiatives are woven into the organisation's culture instead of existing as separate programs. The incorporation of wellness into daily routines via leadership encouragement, inclusive initiatives, and consistency with company values cultivates a collective sense of accountability for well-being among all employees. This integration prompts employees to see wellness not only as a benefit but also as an essential component of their professional and personal development. This fosters an encouraging atmosphere where positive habits, candid discussions about mental and physical wellness, and shared motivation are customary, ultimately enhancing employee involvement, workplace spirit, and a feeling of inclusion.

9.2.5 Future Directions

I. **Standardized Global Guidelines**

The WHO and the ILO are investigating models for digital health and AI in the workplace. Global standardized guidelines in wellness offer a cohesive framework for organizations to create, execute, and assess workplace well-being initiatives across various sectors and areas. These guidelines, created by global organizations such as the World Health Organization (WHO) and the International Labour Organization (ILO), outline shared principles regarding mental health assistance, stress reduction, healthy workplace settings, and the responsible implementation of digital health technologies [18]. Organisations can achieve consistency, inclusivity, and fairness in wellness initiatives by aligning them with global standards while also adapting to local cultural and legal contexts. These guidelines assist employers in comparing their practices to global best standards while also fostering accountability, international cooperation, and the enduring sustainability of wellness as a crucial element of organizational culture.

II. **AI Ethics and Wellness Certification**

Organizations might soon embrace certifications demonstrating that their AI wellness tools are ethical, secure, and efficient. AI ethics and wellness certification is a new framework aimed at guaranteeing that AI-powered wellness initiatives in workplaces are executed responsibly, openly, and inclusively. These certifications ensure that wellness technologies—including mental health chatbots and predictive analytics systems comply with rigorous standards for data privacy, equity, and non-discrimination. Through certification, organizations show their dedication to protecting employee rights while advancing health and well-being via ethical AI practices [19]. This fosters employee trust and assurance in the use of AI-driven wellness instruments while increasing organizational credibility, adherence to regulatory requirements, and conformity with international ethical benchmarks. Ultimately, AI ethics and wellness certification may serve as a standard for ethical corporate wellness, guaranteeing that advancements in employee welfare are aligned with accountability and human principles.

III. **Policy-Driven Gamified Wellness**

Authorities and businesses might advocate for AI gamification strategies to increase adoption (e.g., tax incentives for wellness participation). Policy-led gamified wellness combines defined organizational policies with game-like features to increase employee involvement in health and wellness programs. Organizations can apply gamification methods—such as points, leaderboards, challenges, and rewards by incorporating policies that promote equitable participation, data protection, and inclusivity to increase engagement and motivation in wellness activities. Well-defined policies guarantee that these gamified initiatives are voluntary, transparent, and encouraging instead of competitive or forceful, thus safeguarding employee trust [20]. When effectively managed, gamified wellness initiatives enhance participation and the adoption of healthy behaviors while also promoting teamwork, camaraderie, and a culture of ongoing progress, ultimately benefiting both personal well-being and organizational success. AI wellness initiatives guided by policy guarantee that the ability of AI to observe, forecast, and improve workplace health is complemented by ethical safeguards, privacy protections, and inclusive structures. They transform wellness from an optional benefit into a mandated duty for the organization.

9.3 Building a Future-Ready AI-Driven Wellness Roadmap

The principles of workplace wellness are subtly being redefined by artificial intelligence. Beyond merely tracking steps or delivering standard reminders, AI is now attuned to the distinctive patterns of each employee's life, actively listening, learning, and responding. Wellness applications are commonly utilized as mobile or desktop resources that help maintain a healthy lifestyle [21]. A combination of attributes is provided to promote physical health, mental equilibrium, and emotional wellness. Customized advice from trainers, nutritionists, and intelligent notifications is offered to maintain user motivation. Typically, the options available include fitness and activity; workouts and steps can be effortlessly monitored, and wearables can be connected to the app for this reason. Nutrition—meals can be arranged in just a few taps, calories can be tracked, and food consumption can be recorded, with all calculations performed automatically by the application. Mental well-being and meditation—this format includes guided sessions, relaxation techniques, and tracking moods.

The value of these applications is acknowledged because of their availability at all times and locations. Motivation can be gained while in transit, and evening physical activities after work can be organized [22]. A progress tracker is consistently presented to show when water should be ingested or how many steps remain to be completed. A dependable wellness partner is fundamentally held in the pocket. This choice is deemed ideal for those beginning a healthy lifestyle or aiming to sustain it. By 2025, the wellness sector is projected to thrive. People are increasingly interested in maintaining their health, prompting businesses to assist them in this endeavour. Consequently, numerous companies, fitness centers, and trainers are creating wellness apps. Helpful tips are offered, progress is monitored, and user connectivity is upheld at all times and locations.

The creation of a wellness app usually starts with a discovery phase. At this point, the groundwork for upcoming progress is established, business objectives are specified, and all essential product specifications are gathered. Market and user analysis, together with any available resources, are taken into account [23]. Trends at present can also be examined if necessary, and concepts can be integrated into the foundation of the wellness app. At this point, the distinctive value of the product is also formed. The plan is subsequently presented. Essential and supplementary features are specified, revenue models are explored, and possibilities for MVP are detailed.

After the scope is defined, the design phase begins. At this point, the product is developed using user flows, wireframes, and prototypes. All that has been discussed is converted into visual forms. The experience designed for users transforms into something tangible and perceptible. The subsequent phase is development. In this stage, the visuals are animated by the team. Every element is made clickable. The frontend is created to encompass all that users see and engage with, such as dashboards, animations, and trackers. The backend is designed to support the app internally, manage data storage, integrate, and user logins, and guarantee smooth and secure operations. Subsequently, testing is carried out. Each screen, button, and feature is scrutinized to verify that the app operates correctly and, crucially, satisfies essential business requirements [24]. After successful testing, the launch phase is executed. An app that is predesigned is available on the App Store and Google Play, featuring optimized assets and descriptions. The phases of the AI-based wellness program plan are outlined below.

9.3.1 Setting Clear Objectives

Wellness objectives for organizations such as minimizing burnout, enhancing mental health, or decreasing healthcare expenses, are established prior to implementing AI solutions. Establishing precise objectives in a roadmap for an AI-based wellness program refers to the creation of distinct, quantifiable, and practical goals that direct the program's design, execution, and assessment. These objectives lay the groundwork for integrating AI capabilities with the comprehensive wellness strategy, guaranteeing that every effort is intentional and focused on results. The objectives must be distinctly outlined, ensuring that there is no uncertainty. For example, rather than a broad aim such as "enhance employee well-being," a specific target would be stated as "lower employee stress levels by 20% in 12 months utilizing AI-driven stress management solutions." Every objective should feature measurable criteria, enabling the monitoring of progress and the assessment of success [25]. Metrics such as engagement levels, health results, or interactions with AI-based wellness tools may be employed. Direct links need to be made between goals and wider organizational priorities, such as increasing productivity, lowering absenteeism, or enhancing employee satisfaction. This alignment guarantees that the AI-powered wellness initiative is seen as a strategic resource instead of an isolated project. Achievability is guaranteed via tangible actions backed by AI technologies, such as tailored coaching, predictive analytics, and automated health monitoring. Deadlines or schedules are established to reach goals to ensure momentum and responsibility.

9.3.2 Adopt Ethical and Policy Frameworks

Align programs with international regulations (e.g., the EU AI Act, NDHM, HIPAA) to guarantee data protection, equity, and ethical application of AI. Implementing ethical and policy frameworks in an AI-based wellness program roadmap entails creating and following a collection of principles, guidelines, and rules that regulate the responsible creation, implementation, and utilization of AI technologies within the wellness project. This entails protecting user privacy, maintaining data security, encouraging transparency, and avoiding algorithmic bias or discrimination [26]. Incorporating ethical considerations and adhering to relevant legal standards (such as the GDPR or HIPAA) into the program's core allows organizations to foster trust with participants, safeguard sensitive health information, and guarantee that AI-driven wellness solutions are equitable, responsible, and consistent with both organizational principles and societal norms.

9.3.3 Integrate Predictive Analytics

AI can be used to detect initial indicators of stress, exhaustion, and ongoing illnesses, allowing proactive interventions instead of responsive actions. The incorporation of predictive analytics into an AI-based wellness program framework requires the use of sophisticated data modelling and machine learning methods to foresee upcoming health patterns, threats, and participant necessities [27]. Through the examination of past and current data including biometric details, lifestyle behaviors, and interaction trends—the program can proactively recognize individuals susceptible to specific conditions, suggest tailored interventions, and enhance resource distribution.

9.3.4 Ensuring Inclusivity and Accessibility

Design AI wellness platforms that cater to diverse employee needs, with multilingual, disability-friendly, and culturally adaptive features. *Ensuring Inclusivity and Accessibility* in an AI-driven wellness program roadmap emphasizes designing solutions that are equitable, user friendly, and accessible to individuals of diverse backgrounds, abilities, and needs. This ensures that wellness tools powered by AI are free from bias, culturally sensitive, and adaptable to different languages, literacy levels, and socioeconomic contexts [28]. Accessibility features such as assistive technologies, multimodal interfaces, and inclusive design principles allow people with disabilities or limitations to participate fully. By prioritizing inclusivity, organizations foster fairness, broaden engagement, and create a wellness ecosystem where every individual—regardless of their physical, cognitive, or cultural differences can benefit equally from AI-driven health and well-being initiatives.

9.3.5 Leveraging Gamification and Engagement Tools

AI-driven nudges, challenges, and incentives are utilized to create engaging and motivating experiences in wellness programs. “Incorporate gamification and engagement tools” within an AI-focused wellness strategy highlights the application of game-like tactics and interactive technologies to increase involvement, enthusiasm, and sustained dedication to health and wellness efforts [29]. When features such as rewards, challenges, leaderboards, progress monitoring, and tailored feedback are incorporated, gamification turns ordinary wellness tasks into captivating and enjoyable activities. AI enhances this method by customizing challenges, adjusting difficulty, and providing real-time feedback that maintains user motivation.

9.3.6 Enabling Human-AI Collaboration

AI-generated insights are merged with human supervision via HR, counselors, and managers to ensure empathy and trust in interventions. “Facilitating human–AI cooperation” in an AI-based wellness program strategy underscores the collaboration between human knowledge and AI-driven insights to achieve more tailored, ethical, and impactful wellness results. Instead of being a substitute for human participation, AI serves as a smart support system that evaluates data, recognizes trends, and offers suggestions. Simultaneously, healthcare practitioners, wellness advisors, or organizational leaders contribute empathy, contextual awareness, and ethical reasoning to the decision-making process [30]. This partnership guarantees that wellness initiatives remain centered on people, balancing technological accuracy with emotional understanding and building trust. By facilitating collaboration between humans and AI, organizations can develop a more comprehensive, adaptable, and effective wellness ecosystem.

9.3.7 Monitor, Measure and Improve

KPIs (engagement, health results, absentee rates, productivity) should be set, and wellness initiatives should be consistently enhanced with AI-powered dashboards. “Track, Assess and Enhance” in an AI-driven wellness program strategy emphasizes creating an ongoing feedback system where wellness efforts are monitored, analysed, and improved to optimize their effectiveness. AI tools allow for immediate tracking of health metrics, participation rates, and program results, whereas data analysis offers quantifiable understandings of efficiency and opportunities for enhancement [31]. By establishing explicit metrics, frequently evaluating progress, and modifying strategies on the basis of data, organizations guarantee that wellness programs remain active, pertinent, and outcome focused. This repetitive method increases efficiency while promoting sustainable, ongoing improvements in employee welfare and organisational vitality [21].

9.3.8 Foster a Wellness Culture

Integrate AI wellness tools into the organization's culture via leadership support, communication strategies, and sustained dedication. Promote a "Wellness Culture" within an AI-centric wellness program roadmap, focusing on establishing a setting where health, well-being, and harmony are integral to organizational values, practices, and everyday engagements [32, 33]. In addition to technology and tools, it encompasses fostering a nurturing attitude that promotes openness, inclusiveness, and active engagement in wellness programs. AI can significantly contribute by offering tailored suggestions, monitoring behavioural patterns, and pinpointing areas that require cultural reinforcement.

9.4 Conclusions

Global best practices in AI-enhanced workplace wellness highlight the fusion of technology with a human-centric strategy to foster healthier, more efficient, and inclusive workplace settings. Organizations around the globe are implementing AI solutions that tailor wellness plans, enhance mental health, and foster physical and emotional well-being while prioritizing inclusivity, accessibility, and data protection. The most efficient strategies include ongoing evaluation, responsible AI utilization, and coherence with company objectives, leading to quantifiable enhancements in employee contentment, retention, and overall performance. Through promoting a wellness-oriented culture, utilizing gamification, and facilitating human–AI teamwork, organizations establish standards for sustainable and scalable wellness programs. Ultimately, embracing these best practices indicates a move towards proactive, data-informed, and internationally aligned strategies that place employee well-being at the center of organizational success.

References

1. Li, M. L. Y., Min, S. L. S., & Sündermann, O. (2024). Efficacy of the mHealth app intellect in improving subclinical obsessive-compulsive disorder in university students: Randomized controlled trial with a 4-week follow-up. *JMIR mHealth and uHealth, 12*(1), e63316.
2. Yokomitsu, K., Oimatsu, R., Toh, S. H. Y., & Sündermann, O. (2025). Assessing the efficacy of the INTELLECT cognitive behavioral therapy mobile app for anxiety and depressive symptoms

among at-risk Japanese employees: Randomized controlled trial. *JMIR mHealth and uHealth, 13*, e60871.
3. Piispanen, J. R., & Rousi, R. (2024). Emotion AI in workplace environments: A case study. In *International conference on software business* (pp. 142–148). Springer.
4. Liu, Y., Kauttonen, J., Zhao, B., Li, X., & Peng, W. (2024). Towards emotion AI to next generation healthcare and education. *Frontiers in Psychology, 15*, 1533053.
5. Feng, T., et al. (2025). TILES-2018 sleep benchmark dataset: A longitudinal wearable sleep data set of hospital workers for modelling and understanding sleep behaviors. arXiv preprint arXiv:2507.03520
6. Mundnich, K., Booth, B. M., Lhommedieu, M., Feng, T., Girault, B., Lhommedieu, J., et al. (2020). TILES-2018, a longitudinal physiologic and behavioral dataset of hospital workers. *Scientific Data, 7*(1), 354.
7. Health Navigator. https://en.wikipedia.org/wiki/Health_Navigator?utm_source=chatgpt.com. Accessed 4 September 2025
8. Ekandjo, T. (2024). *Human–AI interaction in regulating productivity and wellbeing* (Doctoral dissertation, Open Access Te Herenga Waka-Victoria University of Wellington).
9. Van Der Maden, W., Lomas, D., & Hekkert, P. (2023). A framework for designing AI systems that support community wellbeing. *Frontiers in Psychology, 13*, 1011883.
10. Ghosh, S., Majumder, S., & Peng, S. L. (2023). An empirical study on adoption of artificial intelligence in human resource management. In *Artificial intelligence techniques in human resource management* (pp. 29–85). Apple Academic Press.
11. Al-Azzawi, A. (2021). AI and well-being: Can AI make you happy in the city. In *Artificial intelligence in the gulf: Challenges and opportunities* (pp. 163–201). Springer.
12. Shin, Y. (2025). Toward human-centered artificial intelligence for users' digital well-being: Systematic review, synthesis, and future directions. *JMIR Human Factors, 12*(1), e69533.
13. Tahvanainen, L., Tetri, B., & Ahonen, O. (2024). Exploring and extending human-centered design to develop AI-enabled wellbeing technology in healthcare. In *Nordic conference on digital health and wireless solutions* (pp. 288–306). Springer.
14. Verma, S., Jain, A., Shukla, S., & Kothari, V. (2024). AI driven medical prediction and suggestive system to establish good health and well-being in society: A real time case study. In *International conference on intelligent and fuzzy systems* (pp. 179–185). Springer.
15. Calvo, R. A., Peters, D., Vold, K., & Ryan, R. M. (2020). Supporting human autonomy in AI systems: A framework for ethical enquiry. In *Ethics of digital well-being: A multidisciplinary approach* (pp. 31–54). Springer.
16. Rosca, C. M., & Stancu, A. (2024). Fusing machine learning and AI to create a framework for employee well-being in the era of industry 5.0. *Applied Sciences, 14*(23), 2076–3417.
17. Montag, C., Rahman, S., Thrul, J., AlBeyahi, F., & Ali, R. (2025). Bringing computing and social/medical sciences together: The case of digital well-being and AI well-being. *Telematics and Informatics Reports, 34*, 100225.
18. Zamani, S., Rostami, A., Nguyen, M., Sinha, R., & Madanian, S. (2025). A gamified AI-driven system for depression monitoring and management. *Applied Sciences, 15*(13), 7088.
19. Tolks, D., Schmidt, J. J., & Kuhn, S. (2024). The role of AI in serious games and gamification for health: Scoping review. *JMIR Serious Games, 12*(1), e48258.
20. https://www.xsolis.com/blog/case-studies-of-successful-implementations-of-ai-in-health care/. Accessed 17 November 2025
21. Mittal, S., Wijethilak, H. P., & Chandel, A. (2025). Effectiveness of AI-powered wellness in promoting employee wellbeing and happiness. In *Insights on wellbeing and happiness in the workplace* (pp. 87–108). IGI Global Scientific Publishing.
22. Butalia, A., Lokhande, K., & Tathod, A. M. (2025). Leveling up health: Unlocking the power of gaming in wellness fusing digital therapeutics. In *International conference on recent advancements and modernisations in sustainable intelligent technologies and applications (RAMSITA 2025)* (pp. 641–653). Atlantis Press.
23. Majumder, S., & Dey, N. (2022). Artificial intelligence and knowledge management. In AI-empowered knowledge management (pp. 85–100). Springer.

24. Ganesan, A. (2025). Transforming wellness: Ethical, cognitive, and privacy challenges in AI-driven healthcare. In *Harnessing AI and machine learning for precision wellness* (pp. 481–500). IGI Global Scientific Publishing.
25. Oluwagbade, E. (2025). Ethical AI in employee wellness: Balancing data-driven insights with employee privacy and trust.
26. Parasa, S. K. (2023). Impact of AI in employee wellness and well-being programs. *European Journal of Advances in Engineering and Technology, 10*(5), 98–100.
27. https://www.vktr.com/ai-disruption/5-ai-case-studies-in-health-care/. Accessed 17 November 2025
28. Narasannagari, L., & Bhardwaj, D. S. (2025). From algorithms to wellness: Uniting AI and healthcare for precision outcomes. In *Harnessing AI and machine learning for precision wellness* (pp. 75–110). IGI Global Scientific Publishing.
29. Dhinakaran, D., Raja, S. E., Jasmine, J. J., Kumar, P. V., & Ramani, R. (2025). The future of well-being: AI-powered health management with privacy at its core. *Wellness Management Powered by AI Technologies, 48*, 363–402.
30. Laffier, J., Rehman, A., & Westley, M. (2025). A conceptual framework to understand the relationships between digital wellness and artificial intelligence. *Cyberpsychology, Behavior, and Social Networking, 38*, 1524.
31. Sqalli, M. T., Al-Thani, D., Qaraqe, M., & Fernandez-Luque, L. (2021). Perspectives on human-AI interaction applied to health and wellness management: Between milestones and hurdles. In *Multiple perspectives on artificial intelligence in healthcare: Opportunities and challenges* (pp. 41–51). Springer.
32. Saeidnia, H. R., Hashemi-Fotami, S. G., Lund, B., & Ghiasi, N. (2024). Ethical considerations in artificial intelligence interventions for mental health and well-being: Ensuring responsible implementation and impact. *Social Sciences, 13*(7), 381.
33. Sahni, M. (2025). When healing becomes a game: Usage of AI and gamification in healthcare and their ethical and legal challenges. In *Proceedings of the 2025 international conference on artificial intelligence, computer, data sciences and applications (ACDSA)* (pp. 1–5). IEEE.

Chapter 10
Conclusions

Mind–body wellness emphasizes the integral link between mental, emotional, and physical health, stressing that genuine well-being can be realized only when all aspects are cultivated in balance. Activities such as mindfulness, meditation, balanced eating, regular exercise, and stress management form a comprehensive approach that enhances resilience, concentration, and overall well-being. The incorporation of mind–body techniques into personal habits, work environments, and healthcare frameworks enables individuals to achieve greater productivity, enhanced emotional well-being, and improved long-term wellness. Ultimately, promoting mind–body wellness is not solely about avoiding disease—it involves cultivating a lasting state of inner equilibrium, energy, and overall well-being. AI for work–life balance highlights the ability of technology to promote a healthier balance between job obligations and personal well-being. Mind–body wellness emphasizes the unbreakable bond between mental, emotional, and physical health, underlining that genuine well-being is attainable only when all aspects are cultivated in balance. Mindfulness, meditation, balanced nutrition, consistent exercise, and stress management form a comprehensive approach that enhances resilience, concentration, and overall quality of life. The incorporation of mind–body techniques into personal habits, work environments, and healthcare systems enables individuals to achieve improved productivity, emotional well-being, and enhanced long-term wellness. Ultimately, promoting mind–body wellness extends beyond merely avoiding illness; it involves cultivating a lasting sense of inner harmony, energy, and overall well-being. AI for work-life harmony highlights technology's capacity to promote a healthier balance between work responsibilities and personal well-being. By leveraging data-driven insights, personalized interventions, and real-time monitoring, AI enables tailored wellness experiences that address both physical and mental well-being. These programs not only help reduce stress, absenteeism, and healthcare costs but also foster a culture of balance, inclusivity, and continuous improvement. When implemented ethically and aligned with organizational goals, AI becomes a strategic enabler of sustainable workplace wellness, ensuring that both employees and businesses thrive in an evolving world of

S. Majumder and N. Dey, *AI Driven Wellbeing to Enhance Lives and Work Environments*, Studies in Computational Intelligence 1257,
https://doi.org/10.1007/978-3-032-15735-5_10

work. AI applications for general well-being demonstrate how technology can act as a powerful partner in promoting healthier, more balanced, and fulfilling lives. From personalized fitness and nutrition guidance to mental health support, stress management, and preventive healthcare, AI enhances the accessibility and effectiveness of wellness solutions for diverse populations. Through the use of data-informed insights, individualized interventions, and immediate oversight, AI facilitates customized wellness experiences that cater to both physical and mental health. These initiatives not only help lower stress, absenteeism, and healthcare expenses but also promote a culture of balance, inclusivity, and ongoing improvement. When deployed ethically and in harmony with company objectives, AI serves as a vital facilitator of sustainable workplace wellness, allowing employees and organizations to flourish in a changing work environment. AI solutions designed for overall well-being illustrate how technology can be a significant supporter in fostering healthier, more balanced, and rewarding lives. AI improves the accessibility and effectiveness of wellness solutions for various populations, offering tailored fitness and nutrition advice, mental health assistance, stress management resources, and preventive healthcare. Through the assessment of personal requirements and the provision of customized support, AI promotes proactive self-care, enabling individuals to make informed decisions about their lifestyles. AI is becoming a powerful influence in developing an organizational wellness culture by providing data-driven insights, tailored interventions, and proactive support systems that meet the varied needs of employees. In addition to enhancing personal health outcomes, AI fosters an environment of compassion, inclusiveness, and balance where wellness is integrated into everyday practices and corporate principles. When applied ethically and strategically, AI not only alleviates stress and burnout but also increases engagement, collaboration, and overall productivity. The incorporation of AI in workplace wellness culture signifies a transition from reactive wellness programs to a sustainable, forward-thinking approach that views employee well-being as fundamental to organizational achievement. Ethical issues and challenges highlight the necessity of accountability, equity, and openness in the creation and use of contemporary technologies and organizational methods. Concerns such as data privacy, bias in algorithms, accountability, and inclusiveness need to be thoughtfully considered to safeguard individuals while fostering trust and fairness. Inadequate governance can compromise both effectiveness and credibility, resulting in ethical failures that pose risks to organizations and society. Consequently, the way ahead necessitates harmony between innovation and responsibility, with ethical guidelines, adherence to regulations, and ongoing assessment steering decision-making. Integrating ethics throughout all phases of strategy and execution allows organizations to convert challenges into chances for creating sustainable, reliable, and human-focused results. Emerging trends in AI for comprehensive wellness indicate a shift towards a more tailored, anticipatory, and unified strategy for human health. Progress in predictive analytics, wearable devices, and emotional AI will enable people to obtain immediate insights and support customized to their physical, mental, and emotional requirements. AI tools that integrate inclusivity, accessibility, and ethical protection will increasingly enhance workplace wellness programs, healthcare systems, and everyday lifestyle management. With the advancement of

AI, attention will transition from merely addressing illnesses to promoting enduring balance and resilience, establishing holistic wellness as a common goal for individuals, organizations, and society. In the end, the future of AI in wellness depends on developing a human-focused system where technology enhances healthier and more satisfying lives.

GPSR Compliance

The European Union's (EU) General Product Safety Regulation (GPSR) is a set of rules that requires consumer products to be safe and our obligations to ensure this.

If you have any concerns about our products, you can contact us on ProductSafety@springernature.com

In case Publisher is established outside the EU, the EU authorized representative is:

Springer Nature Customer Service Center GmbH
Europaplatz 3
69115 Heidelberg, Germany

Batch number: 10755314

Printed by Printforce, the Netherlands